Intergovernmental Fiscal Transfers, Forest Conservation and Climate Change

Intergovernmental Fiscal Transfers, Forest Conservation and Climate Change

Silvia Irawan

Earth Innovation Institute, USA

Luca Tacconi

The Australian National University, Canberra, Australia

Cheltenham, UK • Northampton, MA, USA

Published by
Edward Elgar Publishing Limited
The Lypiatts
15 Lansdown Road
Cheltenham
Glos GL50 2JA
UK

Edward Elgar Publishing, Inc.
William Pratt House
9 Dewey Court
Northampton
Massachusetts 01060
USA

A catalogue record for this book
is available from the British Library

Library of Congress Control Number: 2015952672

This book is available electronically in the **Elgar**online
Social and Political Science subject collection
DOI 10.4337/9781784716608

ISBN 978 1 78471 659 2 (cased)
ISBN 978 1 78471 660 8 (eBook)

Typeset by Servis Filmsetting Ltd, Stockport, Cheshire
Printed and bound in Great Britain by TJ International Ltd, Padstow

Contents

Figures

Tables

Acknowledgements

Research for this book was funded by the Centre for International Agricultural Research (ACIAR) project FST/2007/52 'Improving governance, policy and institutional arrangements to reduce emissions from deforestation and degradation (REDD)', and the Australian Leadership Awards (ALA) scholarship for Silvia Irawan.

We would like to thank the ACIAR project team members, Fitri Nurfatriani, Silahuddin, John Moesiri, Herbert and Zahrul Muttaqin for their support for fieldwork in Riau and Papua. Colleagues from the Ministry of Forestry, the Ministry of Finance and the Ministry of Development Planning, who cannot be named one by one, also provided significant support for fieldwork. We would also like to thank our colleague Irene Ring, who authored with us the papers that are included in this book (in a slightly edited format) as Chapters 7 and 8. Finally, we would like to thank Chris Ulyatt for his painstaking editorial assistance.

Acknowledgements

1. Introduction

Properly designed intergovernmental fiscal transfers (IFTs) present an innovative instrument that creates incentives for public actors to support conservation. Whilst much of the literature on payments for environmental services has focused on private actors, such as smallholders and companies, and their forgone benefits due to conservation (Wunder, 2007), the economic implications of conservation for public actors have received less attention. Yet, this group of stakeholders should not be overlooked. When the state claims ownership of land – a situation common in many tropical forest-rich countries (Tacconi et al., 2010) – public actors are responsible for maximising the revenues from resource utilisation. This does not imply that revenue maximisation is, or should be, the only parameter used by governments to make decisions concerning natural resources. There is evidence, however, that it is a significant determinant of resource management, as noted, for example by Barr et al. (2006) and Andersson et al. (2006). Conservation, however, restricts public actors from generating public revenues, as they can no longer issue permits to pursue income-generating activities in areas designated for conservation. Local governments usually obtain a share of revenues from those resources. As these revenues can no longer be generated due to conservation, compensation to reconcile local costs with the benefits that reach beyond local boundaries is required for local governments to support conservation.

IFTs that include ecological indicators for the allocation of transfers to decentralised governments, and are therefore also known as ecological fiscal transfers, have been advocated as a means to address the spillover effects of the provision of environmental services. This provision of environmental goods and services, such as pollution control and conservation, create spillover effects (Oates, 2001; Sigman, 2005; Kunce and Shogren, 2005; Ring et al., 2010). These effects, also referred to as spatial externalities, occur when public-service provision generates benefits that reach beyond the boundaries of the administrative unit generating the service, while the costs are borne only by the local residents of the unit and the unit itself (Oates, 1972). Due to these externalities, local governments often neglect the benefits that spill over outside their administrative boundaries during the decision-making process. As a result, they tend to provide services below the efficient level (Oates, 1972). To address the spatial externalities of conservation, ecological fiscal transfers have been implemented in Brazil and Portugal and proposed in Switzerland, Germany and India to compensate the forgone opportunity costs of protected areas (Grieg-Gran, 2000; May et al., 2002; Köllner et al., 2002; Ring, 2002; Ring, 2008b; Ring, 2008c; Kumar and Managi, 2009; Santos et al., 2012). In Brazil and Portugal, the transfers have been used to distribute a portion of national governments' taxes (for example the value-added tax) to the states, based on a set of conservation and ecological indicators (Grieg-Gran, 2000; May et al., 2002; Ring, 2008c; Santos et al., 2012). The State of Pará in Brazil has adopted new legislation, as part of its Green Municipality program, where the state government will provide revenues for municipalities for their progress toward reducing deforestation in addition to their effort to support the existing protected areas.[1]

International experience with IFTs shows that funding is mostly sourced from domestic public finance, mainly national government taxes. As the global community becomes increasingly interested in paying for environmental services, international finance may become available to support conservation at the subnational level, and could be channelled through an IFT mechanism. For instance, the agreement on Reducing Emissions from Deforestation and Forest Degradation (REDD+), which was reached at the 2010 Cancun meeting of the UN Framework Convention on Climate Change (UNFCCC), is expected to result in a flow of funds from developed to developing countries. Public finance that has been mobilised (and approved) to support REDD+ between 2008 and November 2011 has been estimated at $446 million, of which $252 million has been disbursed (Nakhooda et al., 2011). REDD+ finance is currently 13 per cent of total climate finance. At the end of 2013, the Conference of the Parties of the United Nations Framework Convention on Climate Change adopted a decision that provides guidance for ensuring environmental integrity for the full implementation of REDD+ activities on the ground. The package also provides a foundation for transparency and integrity of REDD+

action, clarifies ways to finance relevant activities and how to improve coordination of support (UNFCCC, 2013).

A REDD+ scheme would require developing countries to set aside additional forest conservation areas, which may not necessarily yield additional environmental services for local residents. Additional forest conservation would compete with other land-use activities, such as commercial logging, timber plantations and oil palm plantations, which can legally take place in productive forestlands. REDD+ measures that impose restrictions on the development of those land-use activities could therefore lead to a substantial loss of public revenues at the various government levels. At the subnational level, it can be expected that local governments would be interested in REDD+ only if the benefits and costs of conservation were duly acknowledged.

This book builds on the opportunity presented by the global interest in REDD+, and uses it as a case study to examine key design aspects of IFTs for conservation in developing countries. It explores the potential to use IFTs as a means to distribute payments for conservation between different levels of government. The distribution of international payments through IFTs may require a new approach to the design of IFTs, as their purpose is not only to correct spatial externalities of public-service provision (conservation) but also to distribute public revenues vertically between government levels from resource utilisation.

1.1 THE IMPORTANCE OF IFTs FOR DECENTRALISED FOREST MANAGEMENT

Developing countries have progressed towards decentralising functions for the provision of basic public services to subnational governments. A similar trend can also be observed in forest management, where powers to manage forest resources are being transferred to subnational governments (Larson and Soto, 2008; Larson, 2003; Ribot et al., 2006). Decentralisation is advocated on the basis of bringing the decision-making process closer to the public to ensure that policies better suit local needs (Cheema and Rondinelli, 1983; Conyers, 1983; Rondinelli, 1990). Decentralisation in public administration also needs to be accompanied by the transfer of authority to subnational governments to generate public revenues to finance local public services. Through fiscal decentralisation, national governments can increase the tax authority of subnational governments or increase transfers to local governments to finance public-service provision at that level (Schneider, 2003; de Mello, 2000; Bräutigam, 2002; Falleti, 2005). However, despite the massive decentralisation process in public administration, national governments in developing countries are often not willing to pursue real fiscal decentralisation as they fear that subnational governments cannot deliver the public services devolved due to, *inter alia*, the low capacity of subnational governments (Bahl and Wallace, 2005).

Decentralisation in forest management transfers administrative functions to subnational governments to deliver a number of forest-related services, including forest monitoring and illegal logging control (Andersson et al., 2004; Andersson et al., 2006; Larson, 2003). However, as forest resources involve strong economic and political dimensions (Larson, 2003), decentralised forest management is often complicated by a lack of meaningful powers and sufficient resources being transferred to local authorities. National governments often refuse to devolve authority over raising or spending revenues and deciding the utilisation of high value resources to subnational governments (Ribot et al., 2006). Local governments, in turn, feel that they have been given the burden of delivering forest services without having the income to manage forests (Larson, 2003). When administrative functions are transferred to subnational governments without sufficient resources to fulfil these functions, subnational governments face a mismatch between revenues and expenditures, which will eventually compromise the quality and quantity of service provision (Bahl and Wallace, 2007; Bird and Smart, 2002; de Mello, 2000). The resistance of national governments to pursue true decentralisation policies has compromised the results of decentralised forest management in developing countries (Larson and Soto, 2008; Larson, 2003; Ribot et al., 2006). Moreover, the issue of spillover benefits resulting from forest conservation and protected areas has complicated forest management in developing countries (Ring et al., 2010; Kumar and Managi, 2009; Ring, 2008c)

To address both the mismatch between revenues and expenditures as well as the spatial externalities of service provision, the literature on fiscal decentralisation suggests IFTs as a suitable option. Two important purposes of IFTs are to distribute part of a national government's revenues to close the gap between spending and revenues mobilised locally and to correct spillover benefits of public services to other jurisdictions (Bird and Smart, 2002; Bahl, 2000; Bahl, 1999; Bird, 2001; Bird, 1999). In decentralised countries, it is common practice for national governments to collect most public revenues and leave local governments with a limited tax base (Bräutigam, 2002; de Mello, 2000; Bahl and Wallace, 2007; Bahl and Wallace, 2005). Hence, if subnational governments are considered important providers of certain public goods and services, the higher level government needs to use IFTs to distribute part of its revenues to close the gap between spending and revenues mobilised by the local governments (de Mello, 2000; Bahl and Wallace, 2007; Bird and Smart, 2002). In relation to the spillover effects of public-service provision, by providing a unit of subsidy to local units generating spillover benefits, IFTs can encourage local decision makers to take into account the benefits of public-service provision that spill outside their administrative jurisdictions (Oates, 1972; Bird and Smart, 2002; King, 1984).

Designing IFTs to support conservation in decentralised countries requires a comprehensive examination of the impacts of forest management on the expenditures and revenues of local governments. In forest management, government stakeholders are concerned about providing forest-related services and most importantly about generating public revenues from forest resources. As conservation restricts local governments from generating public revenues from forests, IFTs for conservation are imperative to correct spatial externalities of forest-related services and also to distribute public revenues generated from forest resources between government levels. The distribution of revenues from forests will impact on the capacity of local governments to finance local public-service provision in other sectors, such as health and education. This study therefore aims to design IFTs for conservation through a careful examination of both the expenditure and revenue streams of local governments.

1.2 THE BOOK'S OBJECTIVES AND METHODOLOGY

In developing countries, designing IFTs for conservation requires careful consideration of technical challenges and constraints related to the capacity of implementing agencies. Research on IFTs for biodiversity conservation to date has focused on the distribution formula to estimate the amount of transfers allocated to each unit of subnational government (for example Köllner et al., 2002; Ring, 2008b; Kumar and Managi, 2009). Two other components of the design of IFTs – conditionality and accountability – have received inadequate attention and are therefore considered in this book. Finally, using the case of REDD+ revenue distribution from the national to local governments, we provide a practical example of the design of an IFT for conservation in Indonesia.

The book is organised into three parts. First, it sets the scene by addressing the role of financial incentives and other factors in influencing local governments' interest in, and commitment to, conservation in Indonesia. Second, it addresses the design of IFTs for conservation – focusing on Indonesia's experience as the country case study – with a focus on the distribution formula, conditionality and accountability. Lastly, it provides the practical example of designing an IFT for REDD+ revenue distribution to local governments in Indonesia.

1.2.1 Setting the Scene: The Role of Financial Incentives

The first part sets the scene to understand the role that IFTs can play to support conservation. A review of the literature about decentralised forest management is provided as an overarching theoretical background. This literature has mostly analysed the administrative and political decentralisation implications of forest management (Ribot et al., 2006; Larson, 2003; Tacconi, 2007; Andersson et al., 2004; Andersson, 2004; Andersson and Gibson, 2007; Andersson, 2003). Studies of decentralised forest management have focused specifically on the institutional configurations and balance of power and interactions between actors involved in, or affected by, forest management (Larson and Soto, 2008). Although fiscal incentives have been highlighted as an important condition of success, specific research on decentralised forest management and fiscal decentralisation is lacking. The literature on fiscal decentralisation suggests a number of instruments for the efficient provision of public services at the local level (Bird and Ebel, 2007; Bird and Smart, 2002; Bahl and Wallace, 2007). This part brings together, therefore, the issues of fiscal decentralisation and decentralised forest management to understand how to design instruments that provide incentives for subnational governments to better manage forest resources.

There have been several recent assessments of the extent of deforestation in Indonesia. The government had estimated deforestation over the period 2000–2005 and 2006–2009 at 1.08 and 0.83 million hectares per year (Ministry of Forestry, 2008b; Ministry of Forestry, 2013). A more recent government assessment reported deforestation over the period 2009–2011 at 0.40 million hectares per year (Ministry of Forestry, 2013). However, a more recent, independent peer-reviewed study provides significantly higher estimates of deforestation in recent years, and a very concerning picture in terms of the trend of the deforestation rate. Annual deforestation stood at about 0.21 million hectares in 2001, and it had increased to about half a million hectares by 2005 (Margono et al., 2014). It also stood at around half a million hectares per year during the period 2006–2010, with a peak of about 0.71 million hectares in 2009, but in the last two years of the assessment it showed an upswing, reaching about 0.84 million hectares in 2012 (Margono et al., 2014). These levels of deforestation and forest degradation have positioned Indonesia as the third largest emitter of greenhouse gases (GHG) (World Bank, 2007a). Moreover, deforestation also affects biodiversity, and Indonesia is one of the globe's biodiversity hotspots (Myers et al., 2000). Indonesia's former President Yudhoyono pledged to cut emissions by 26 per cent by 2020 compared to business as usual. The forestry sector accounts for 87 per cent of the total 26 per cent target (Government of Indonesia, 2011). In response to REDD+ negotiations, the government of Indonesia is developing the regulatory and institutional architecture to implement REDD+, using a nationally based approach. Local governments would be encouraged to participate to implement subnational REDD+ projects. Carbon credits generated at the local level would be standardised and accounted for within the national system. This approach requires the national government to distribute the revenues that would be generated from REDD+ to subnational stakeholders.[2]

The alarming rate of deforestation in Indonesia, which is a major national and global concern, and the fact that the country has embarked on a programme to implement REDD+ are the reasons behind the choice of the country for the case study. Two provinces were selected as the focus of this book. They represent diverse cases in terms of deforestation trends, fiscal capacity and special autonomy status. Chapter 3 provides a detailed discussion of the key features of the case study.

Deforestation can be perceived as either good or bad by local stakeholders, including local communities and government officials, since legal land-use changes and forest exploitation generate public revenues. The second half of the first part of the book seeks to understand how financial incentives and other factors influence local governments' interest in reducing deforestation in Indonesia based on their perspectives towards deforestation and forest conservation (Chapter 4). This analysis is important as financial incentives alone may not be sufficient to shift local governments' interest, and may need to be accompanied by other factors, such as devolution of authority and pressure from NGOs. Chapter 4 aims to shed light on the causal mechanism of how different factors may influence local governments' interest in, and commitment to, conservation and/or land-use change in their localities. A causal statement is important in developing a public policy to ensure that the policy has an accurate theory of how to bring about change (Sabatier and Mazmanian, 1980; Sabatier, 1988; Sabatier, 1991; Parsons, 1995; Sabatier and Mazmanian, 1979). It is therefore important to provide a possible explanation of how the behaviour of subnational governments as the target group could be influenced by financial incentives to achieve the desired end-state – that is, reduced deforestation.

1.2.2 The Design of IFTs for Conservation

The second part focuses on the design elements of IFTs for conservation. The literature on IFTs for biodiversity conservation suggests that different options can be developed for fiscal transfers based on theoretical justifications, but political processes and community lobbying will influence the final design of IFTs (Ring, 2008b; Köllner et al., 2002). Specifically, Chapter 5 provides a comprehensive assessment of the literature on the design of IFTs as seen from the perspective of the fiscal decentralisation literature, the international experience on IFTs for conservation, and the existing Indonesian IFT mechanism for the forestry sector and land-based activities.

The second half of the second part presents an analysis of government officials' perspectives about the design of an IFT (Chapter 6). Since political negotiations to decide the final design of IFTs involve a multiplicity of actors, it is imperative to understand their values and interests on the design of IFTs. In order to examine those perspectives, an interpretivist policy analysis approach is adopted. Such an approach advances a *practical* conception of reason as an alternative methodological framework to the traditional form of scientific rationality (Fischer, 2003). When conducting interpretive policy analysis, analysts have to become immersed in the beliefs (ideas, values, feeling and meaning) of the participants (Yanow, 2000). Stakeholders' perspectives about the design of IFTs are socially constructed (Guba and Lincoln, 1994). Hence, we did not aim to find a universal truth about this social situation. Rather, we sought to identify details related to the preferences of stakeholders and to combine them in a sensible and coherent manner (Furlong and Marsh, 2010). The research for Chapter 6 involved interviews with government officials. A constant comparison analysis, applying the principles suggested by the grounded theory method (Robson, 2002; Grbich, 2007), was applied during the analysis of the qualitative data collected. The analysis involves coding and analysing categories and meaning of qualitative data.

1.2.3 Practical Example: Designing IFTs for REDD+ Revenue Distribution

The third part provides a practical example of designing IFTs for conservation using the REDD+ revenue distribution in Indonesia. The description of REDD+ in this specific context is provided in Chapter 8. The two key aspects of IFTs are the determination of grant size and the distribution formula. They are considered in Chapter 9, which looks at those issues with the practical case of allocating REDD+ revenues to subnational governments. To estimate the amount of REDD+ revenue transfers to local governments, the tools used are available from academic disciplines including welfare economics and public finance. But before discussing the grant size and the distribution formula, the existing financial incentive structure that influences subnational governments' decisions needs to be assessed. The incentives affecting land-use change and forest exploitation in Indonesia is therefore considered in Chapter 8, which estimates REDD+ opportunity costs of alternative land uses from the perspective of subnational governments, as well as the central government and companies. That analysis focuses on three land-use activities: commercial logging, timber and oil palm plantations. The opportunity costs of alternative land-use activities can be used as the basis to determine the size of the REDD+ distributable pool allocated to each government level. After the distributable pool is estimated, the study analyses the impact of using different approaches to determine the amount of the transfers for district governments.

NOTES

1. http://policymix.nina.no/News/Newsarticle/tabid/3574/ArticleId/2196/Exciting-development-in-Amazon-ICMS-E.aspx accessed 9 January, 2014.
2. There are three tiers of government in Indonesia: national, provincial and district/municipality level. For simplicity, this study refers to the district/municipality level as the 'district level', as forests are mostly found in districts, while municipalities refer to city areas. Subnational level is used to refer to both provincial and district/municipality levels.

2. Theories of decentralised forest management and fiscal decentralisation

The literature on decentralised forest management has paid significant attention to the administrative and political implications for forest management (Ribot et al., 2006; Larson, 2003; Tacconi, 2007; Andersson et al., 2004; Andersson, 2004; Andersson and Gibson, 2007; Andersson, 2003). Studies of decentralised forest management have focused specifically on the institutional configurations and balance of power and interactions between actors involved in, or affected by, forest management (Larson and Soto, 2008). Although fiscal incentives have been highlighted as an important condition of success, specific research on decentralised forest management and fiscal decentralisation is lacking. The literature on fiscal decentralisation suggests a number of instruments for the efficient provision of public services at the local level (Bird and Ebel, 2007; Bird and Smart, 2002; Bahl and Wallace, 2007). This chapter sets out, therefore, to bring together the issues of fiscal decentralisation and decentralised forest management to understand how to design instruments that provide incentives for subnational governments to better manage forest resources.

Fiscal decentralisation theories are mainly concerned about the assignment of functions to different levels of government and the appropriate fiscal instruments for carrying out these functions (Oates, 1999). They can inform the implementation of decentralised forest management about: the function that should be performed by subnational authorities; how to locate resources generated from forest resources at different governmental levels to optimise social welfare (Musgrave, 1959); and the amount of resources best handled by each government level to deliver forest services (Oates, 1972). Fiscal decentralisation theories also suggest intergovernmental fiscal transfers (IFTs) as a means to ensure the efficient provision of public services at the local level. Fiscal instruments have long been used to address negative externalities of environmental pollution in federal countries (for example Oates, 2001; Sigman, 2005; Kunce and Shogren, 2005) and have recently been used to compensate for local costs of biodiversity conservation resulting from land-use restrictions of forest conservation (Grieg-Gran, 2000; May et al., 2002; Köllner et al., 2002; Ring, 2002; Ring, 2008c; Ring, 2008b; Kumar and Managi, 2009; Ring et al., 2010).

The chapter first discusses administrative, political and fiscal decentralisation as well as how decentralisation in one dimension may influence decentralisation along other dimensions. It then reviews the practice of decentralised forest management in developing countries to understand the factors that determine its successful implementation. The relationship between fiscal decentralisation and decentralised forest management is then discussed. Lessons from the relatively limited body of knowledge on IFTs for biodiversity conservation are considered, together with their potential influence on local governments' interest in supporting forest management and conservation. Specific fiscal instruments that can be used to influence subnational governments' behaviour in land-use management are then presented. Finally, the chapter discusses aspects of the design of IFTs that need to be considered to ensure they achieve their objectives.

2.1 THEORY OF DECENTRALISATION

Decentralisation is defined as the: 'transfer of planning, decision-making, or administrative authority from the central government to local administrative units, semi-autonomous, parastatal organisations, local governments, or non-governmental organisations' (Cheema and Rodinelli, 1983, p. 18). The types of decentralisation are administrative, political and fiscal decentralisation (Falleti, 2005; Schneider, 2003). Administrative decentralisation is the transfer of the administration and delivery of public services such as education, health, or social welfare to subnational governments. Decentralisation of administrative functions may also include the devolution of the decision-making authority and resources to meet the costs of public service deliveries (Falleti, 2005). The transfer of authority to increase the fiscal autonomy of subnational governments, known as fiscal decentralisation, involves increasing fiscal transfers from the central government or transferring tax authority to subnational governments (Schneider, 2003). Political decentralisation is the transfer of political authority or electoral capacities to subnational actors through a set of constitutional amendments and electoral reforms (Falleti, 2005).

Administrative decentralisation involves various degrees of autonomy. The degrees of autonomy can be categorised as deconcentration, delegation and devolution (Rondinelli, 1990; Schneider, 2003). Deconcentration occurs when central governments disperse responsibility for a policy to their field offices and retain authority over the local offices. Central governments exercise their authority through the hierarchical channels of the bureaucracy (Schneider, 2003). Delegation transfers policy responsibilities to subnational governments that are not controlled by the central government but remain accountable to it (ibid.). Finally, devolution takes place when the central government allows local governments to exercise power and control over the transferred policy. The national government can only exercise control over subnational governments by threatening to withhold resources or responsibility from them (ibid.).

Proponents of decentralisation suggest both political and economic rationales. From the political and public administration point of view, decentralisation is expected to (Cheema and Rondinelli, 1983):

- increase sensitivity to local needs and ensure that decision makers are more flexible and innovative. Hence, the policies and decisions made should be better tailored to local needs;
- promote greater participation of local people in the planning and implementation of national development;
- increase political stability by harmonising interest between national and local levels;
- increase the capacity of local governments, especially when the transfer of powers and authorities is followed by adequate transfers of resources.

The economic rationale of decentralisation is to enable local governments to provide public services according to the different preferences of individuals in their jurisdictions. Decentralisation allows individuals to seek out a community that is best suited to their preferences and prevents welfare losses to society caused by the uniform provision of public services (Tiebout, 1956; Oates, 1972). In order to achieve the economic objectives, decentralisation in public administration is usually followed by the devolution of fiscal power from the national government to the subnational governments (Davoodi and Zou, 1998; Bahl, 1999).

Decentralisation along one dimension may influence or cross over into decentralisation in other dimensions. Schneider (2003) suggests that increasing decentralisation in one dimension can lead to an increase in decentralisation in another dimension, however, it could also lead to a decrease in another dimension. Fiscal decentralisation might result in greater administrative decentralisation if subnational governments spend the increased resources to assert their administrative autonomy. However, fiscal decentralisation might lead to a lower degree of administrative decentralisation if national governments increase bureaucratic or regulatory controls as the counteraction of the release of resources (Schneider, 2003). Falleti (2005), who studied the decentralisation process in Latin America, argues that different types of decentralisation are negotiated and enacted at different points in time. When decentralisation is instigated due to subnational pressure, political decentralisation is likely to occur first, then governors and mayors would most probably demand fiscal decentralisation; and finally administrative decentralisation would follow (Falleti, 2005). In contrast, when decentralisation is pursued due to national interest, administrative decentralisation is likely to occur first. If fiscal resources do not accompany the transfer of responsibilities, the national government will strengthen its power by making subnational officials more dependent on transfers from the national government. Following this process, political decentralisation, if it happens, will be the third type of reform (ibid.).

2.2 THEORY OF DECENTRALISED FOREST MANAGEMENT

The rationale for decentralised forest management includes reducing costs, increasing forest department revenues and/or increasing control over local communities (Larson and Soto, 2008). From the administrative decentralisation perspective, the degree of decentralised forest management can also range from deconcentration to devolution. Deconcentration in forest management occurs when the national government provides forest-related services at the local level through its field offices. Delegation in forest management involves the transfer of administrative functions (obligations) to subnational governments to deliver forest services such as: training for local user groups, forest monitoring, controlling illegal logging, preventing and controlling fire, environmental education as well as forest protection and conservation (Gibson and Lehoucq, 2003; Andersson, 2003; Larson 2003; Andersson et al., 2006). In contrast, devolution in forest management transfers power to decide on the utilisation of forest resources to generate public revenues (Ribot et al., 2006; Larson, 2003; Andersson et al., 2006). Decentralisation in forest management often involves only the transfer of administrative functions to provide forest services without devolving the power to generate income from forest utilisation to local governments. Since forest resources can be exploited to achieve economic and political objectives, decentralisation in forest management involves high levels of conflict and resistance from central governments to pursue true decentralisation policies (Larson and Soto, 2008; Larson, 2003; Ribot et al., 2006).

There is a contentious debate about whether subnational governments should be provided with more authority in forest management. Devolution of power to local governments is considered necessary to ensure good forest management at the local level because elected local governments are more likely to be downwardly accountable (Ribot et al., 2006; Ribot, 2003). However, more authority to subnational governments does not always lead to better forest management (Tacconi, 2007; Wunder, 2010). Many developing countries have weak representative decision-making processes and local elite captures are common. As a result, vested interest groups can often take advantage of decentralisation for their own benefits (Francis and James, 2003). Moreover, local people, as the electors, may also prefer forest exploitation for their livelihoods instead of conservation (Tacconi, 2007), which puts pressure on local governments to pursue land-use change and forest exploitation.

The political and economic dimensions of forest management often make it more complicated to administer compared with other public service sectors such as health and education (Larson, 2003). Several empirical studies (for example Ribot et al., 2006; Larson, 2002; Andersson et al., 2004; Andersson and Gibson, 2007; Andersson, 2003) of decentralised forest management in developing countries have suggested that a set of important governance factors are required to ensure successful forest management at the local level (Table 2.1). These factors are: sufficient resources, financial incentives, upward accountability, local governments' commitment, discretionary power, demands from NGOs and pressure from local people.

To understand how governance factors influence forest management at the local level, Larson (2003) developed a model of successful decentralised management. She grouped important factors that determine the outcome of decentralised forest management into: legal structure, local decision-making sphere and mediating factors (Figure 2.1). The legal structure determines the types of powers transferred, the accountability relations of the actors receiving powers and the degree to which the powers transferred represent rights or privileges. The local decision-making sphere determines local government initiatives in the forestry sector and is influenced by four variables: local government capacity, power relations, the incentive structure for forest management and environmental and social ideology. The mediating factors lie between the legal structure and the local decision-making sphere and include the role of the central government (CG) and economic incentives.

Most of the studies listed in Table 2.1 focus more on public service delivery in the forestry sector than on the sustainability of forest management (that is preventing deforestation). The studies apply varying measures of good forest management as their dependent variable including: (i) budget allocated, staff assigned and the level of importance assigned to the forestry sector by local governments (Andersson et al., 2006); (ii) local governments' responsiveness to local people (Andersson, 2004); and (iii) local governments' interventions in the forestry sector (Larson, 2002). Two studies (Andersson and Gibson, 2007; Andersson et al., 2010) consider, however, deforestation as the dependent variable. Andersson and Gibson (2007) analyse the effect of local institutions, national policy, socioeconomic context and biophysical conditions on deforestation. They find that biophysical factors are the strongest determinant of both total and permitted deforestation. According to them, unauthorised deforestation is best explained by a combination of factors related to local institutional performance, national policy and biophysical conditions. They find that the level of municipal forest governance performance has a negative and statistically significant

Table 2.1 Important factors for successful decentralised forest management

Factors	Definition/explanation	Supporting literatures
Capacity/ sufficient resources (technical, administrative, and financial)	The ability of local governments to carry out their mandates, including: financial and administrative management, problem-solving, technical skills and the capacity to be democratic leaders (Larson, 2002)	– Andersson *et al.* (2004): studied 50 Bolivian municipalities – Larson (2002): examined 21 municipal governments in Nicaragua – Gibson and Lehoucq (2003): interviewed mayors in 100 municipalities in Guatemala
Financial incentives	The possibility of increasing municipal incomes (Larson, 2002; Andersson et al., 2006)	– Andersson et al. (2006): studied 100 municipal governments in Guatemala and another 100 in Bolivia – Andersson (2003): studied 50 Bolivian municipal governments – May *et al.* (2002): studied the impact of fiscal transfers for forest conservation – Larson (2002): examined 21 municipal governments in Nicaragua
Upward accountability/ central government coercion	Vertical accountability mechanisms to higher governmental levels – providing information and enabling sanctions	– Ribot *et al.* (2006): used a comparative empirical approach in Senegal, Uganda, Nepal, Indonesia, Bolivia, and Nicaragua – Andersson (2003)*; Andersson et al. (2006)*
Local governments' commitment	Commitment to the long-term sustainable use or protection of forest resources (Larson, 2002)	– Larson (2002): examined 21 municipal governments in Nicaragua – Andersson et al. (2006)*

Table 2.1 (continued)

Factors	Definition/explanation	Supporting literatures
Discretionary powers	Secure domain of autonomous decision-making and funding to implement the authorities devolved	– Andersson (2003)*; Ribot (2003)*; Ribot et al. (2006)*
Demands from NGOs	Local pressure or demands from local non-governmental organisations (NGOs) to take environmental initiatives	– Larson (2002)*; Andersson (2003)*; Andersson et al. (2006)*; Gibson and Lehoucq (2003)*
Local people pressure as the electorates	Local pressure or demands from local people or communities to take environmental initiatives	– Andersson (2003)*; Larson (2002)*; Andersson et al. (2006)*; Ribot et al. (2006)*

Note: *See description above.

effect on unauthorised deforestation. They suggest that municipal governments could mitigate pressure to deforest indiscriminately by facilitating local forest users to obtain more secure forest property rights. They consider that forest users with more secure property rights were less likely to convert forests to competing land uses such as agriculture or pasturelands (Andersson and Gibson, 2007).

Andersson et al. (2010) examine the impact of decentralisation (de facto and de jure) on the change of forest cover in municipalities in Guatemala, Peru and Bolivia. Their study finds that de facto decentralisation has a positive and significant effect on forest cover. De facto decentralisation, in their study, represents the capacity of local governments to gather revenues. Hence, localities with more financial autonomy experience less forest loss as they are more likely to be able to implement effective forest conservation policy and can invest relatively more in forest activities (Andersson et al., 2010).

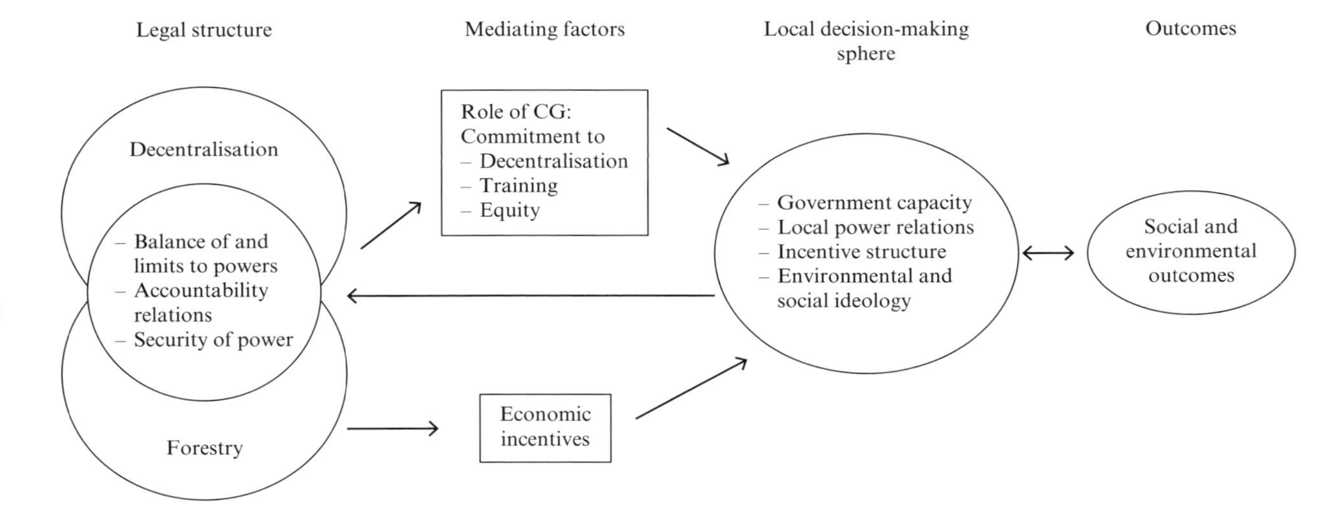

Source: Larson (2003, p. 220)

Figure 2.1 Larson's model of decentralised forest management

They identify the important governance factors affecting local governments' interest in conservation or deforestation as: NGO pressure, local financial importance of forestry; socioeconomic context; national policy and local institutional performance (Andersson et al., 2010; Andersson and Gibson, 2007). These factors, together with biophysical factors, such as access to forests and topography, influence the change of forest covers in municipalities in Latin American countries. Andersson et al. (2010, p. 15) also argue that there is: 'a plausible link between stronger local capacity to raise taxes and more stable forest cover because stronger taxing capacity is associated with greater emphasis on forestry policy'.

It is evident that ensuring local governments manage forests sustainably (or reduce deforestation) presents more challenges than merely having them deliver forestry-related services. Tropical deforestation is a complex event, which varies between regions, driven by a synergy of economic factors, institutions and national policies (Geist and Lambin, 2002). Considering the importance of forest resources for the economy and local livelihoods, Tacconi (2007) proposed additional provisions to Larson's (2003) model (Figure 2.2). In that model, local people influence local government policies, including forest management. Forests can be perceived as a source of livelihoods that could be exploited or preserved to produce environmental services. The model also incorporates three other governance factors into the legal framework: tenure arrangements, corruption and patronage. Finally, economic growth and provision of services may arise from decentralisation processes with possible impacts on forests.

The existing literature on decentralised forest management provides a comprehensive picture about the complexity of local forest governance in developing countries. Adding to the complexity, reducing deforestation and forest degradation may compromise national and local economic growth as well as local livelihoods. Deforestation is determined by national policies, local institutions, biophysical and socio-economic conditions. The national government usually has the authority to make the final decision about policies on permitted or legal deforestation, however the pressure to deforest can also come from local stakeholders. A review of the decentralised forest management literature highlights the important link between financial incentives (devolved through fiscal decentralisation) and deforestation. The next section discusses decentralised forest management

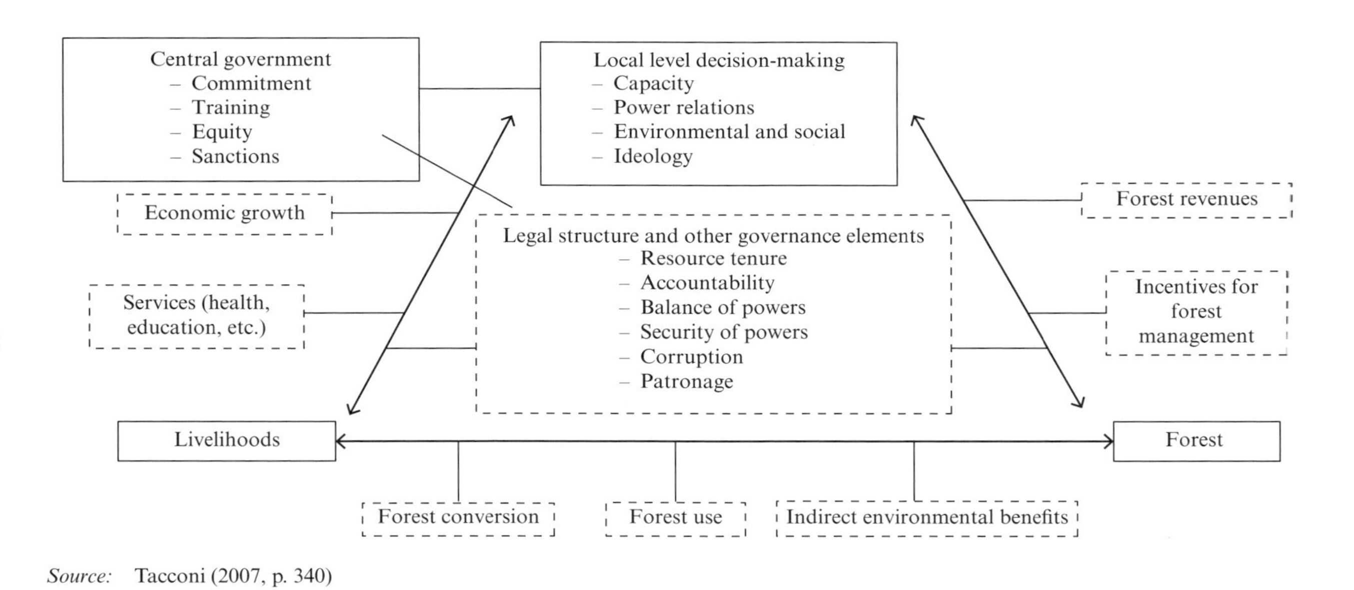

Source: Tacconi (2007, p. 340)

Figure 2.2 Tacconi's model of decentralised forest management

specifically from the fiscal decentralisation perspective, which offers a wide range of suggestions about how to allocate financial resources to different government levels.

2.3 FOREST MANAGEMENT THROUGH THE LENS OF FISCAL DECENTRALISATION

Fiscal decentralisation sets out the roles that the government sector should play. Musgrave (1959, p. 5) proposes that the public sector should play a role in three budget functions: 'the function of providing the satisfaction of public goods and services (allocation); the function of providing for adjustments in the distribution of income (distribution) and the function of contributing to stabilisation (stabilisation).' There is a consensus that the last two functions should be the responsibility of the central government because a local government has: 'a limited scope for effective stabilisation or redistributive policies within its jurisdiction' (Oates, 1972, p. 5). Related to the provision of public services, Oates (1999) contends that public goods that have to be provided to the entire population in the country should be provided by the central government.

Decentralised levels of government are the best level to provide goods and services, where consumption is limited to a particular jurisdiction (Oates, 1999). Local goods and services should be tailored to the preferences and circumstances of the local constituents (Oates, 1999). Musgrave (1959) argues that decentralisation will enable local governments to provide services according to the different circumstances and tastes of individuals in the jurisdictions. This can avoid welfare losses to society caused by the uniform provision of public services in more centralised countries (Oates, 1972). There is, however, no straightforward prescription delineating the specific goods and services to be provided at each level of government. This situation complicates the process of decentralised forest management, where the issue of which functions should be transferred to what level is rather contentious (for example Ribot et al., 2006; Larson, 2003; Larson and Soto, 2008).

Fiscal decentralisation theory also provides suggestions on specific fiscal instruments to underpin the delivery of the functions performed by local governments. In a multilevel public sector, it is important to align responsibilities with fiscal instruments at the proper level of government (Oates, 1999). In order to finance the provision of goods and services, local governments need to be provided with the authority to generate local revenues (Bird and Ebel, 2007). Sources of local governments' revenues are taxes, direct contributions such as charges, public enterprises' profits and royalties from natural resources (Bräutigam, 2002). As the role of local governments is to provide local residents with public services for which they are willing to pay, they should, whenever possible, charge for the services they provide (Bird, 2001, p. 11). When charging is not feasible, services should be financed from taxes collected from the residents. As local governments' tax bases tend to be limited, higher government levels need to share part of their revenues with local governments through intergovernmental fiscal transfers (IFTs) if the latter are important providers of public goods and services (de Mello, 2000).

In most developing and transition countries, IFTs play an important role in subnational government financing (Bahl, 2000). Three of the major purposes of fiscal transfers are (Bird and Smart, 2002; Shah, 2006): (i) distributing part of national government's revenues to close the gap between spending and revenues mobilised locally; (ii) correcting spillover benefits of public services to other jurisdictions; and (iii) ensuring fiscal equality between local governments in a country. A revenue-sharing mechanism usually aims to address the mismatch (fiscal gap) between expenditure needs with public revenues generated at the local level. IFTs can also be implemented to correct spatial externalities that occur when public service provision generates benefits for residents outside a locality. Spatial externalities will create an inefficient outcome as the decision makers often neglect the benefits accrued to outsiders in the decision-making process (Bird, 1999). To address the spatial externalities, local governments should be provided with a unit subsidy, usually in the form of a conditional/unconditional grant, to correct the externalities (Oates, 1972, p. 66). Local governments are then expected to provide the right amount of public service.

From the fiscal decentralisation perspective, forest management has two broad implications: (i) forest resources generate local revenues; and (ii) several forest services (for example forest protection and conservation) need to be performed by subnational governments. The management of forests as resources can impact on the revenue stream, while forest-related services to be provided by local governments have implications for local governments' expenditures.

It is imperative to analyse at what government level revenues from forests should be collected and forest-related services should be performed, before exploring fiscal instruments that are required for successful decentralised forest management.

In decentralised forest management, national governments mostly retain the power to decide on the utilisation of forest resources to generate revenues (Ribot et al., 2006). This arrangement appears to be a common practice in decentralised countries. Searle (2007, p. 393) contends that national governments have greater capacity to: (i) deal with complex royalty schemes for natural resources; (ii) prevent cross-border loss of taxes through firms moving their profit to a lower taxing jurisdiction; and (iii) arrange revenue distribution between jurisdictions. When national governments collect revenues from natural resources, the fiscal decentralisation literature suggests that a portion of revenues collected by the national government should be distributed back to the local governments where they originated (Searle, 2007; Bahl and Wallace, 2007). The distribution of revenues between government levels can use a vertical revenue-sharing mechanism. Governments usually generate public revenues from exhaustible natural resources, including timber extraction, in the form of upfront payments, rent income and back-end payments (Searle, 2007).[1] The purposes of these revenues are to cover public costs associated with any economic activity exploiting natural resources, and/or to fund specific services and/or to add to general revenues to finance a range of services. Subnational governments should obtain a portion of the revenues if they incur expenses from infrastructure responsibilities and other impacts including environmental impact (Searle, 2007). If a local government owns the assets being exploited, then it should receive some additional compensation, although this does not mean other jurisdictions should receive no benefit (ibid.).

To support sustainable forest management and conservation, subnational governments are often required to perform a number of forest services at the local level. The literature on decentralised forest management reports a number of devolved forest-related services, such as:

- training of local user groups and setting up databases on municipal forest resources in Bolivia (Andersson, 2003);
- forest monitoring, illegal logging control, legal logging

supervision, and developing forestry plans with popular participation in Guatemala (Gibson and Lehoucq, 2003; Larson, 2003);
- establishing municipal parks, organising campaigns for fire prevention and control, participating in the management of national parks and promoting citizen participation in Nicaragua (Larson, 2003);
- preventing and controlling fire, environmental education, and promoting agroforestry in Brazil (Larson, 2003).

Subnational governments have an advantage in administering all these services because of their proximity to forest areas and local residents in surrounding forests. If a country devolves the administrative functions (obligations) to perform forest-related services, the process should also be accompanied by adequate financial and administrative resources, otherwise decentralisation efforts will fail when the resources are insufficient (Andersson et al., 2006; Larson, 2002; Larson, 2003). Larson (2003) found that local governments in Latin America often feel that they have been given the burden without the benefits of management – either in terms of the authority or income.

On the basis of this review, there are four possible relationships between the revenue collection and expenditures assigned to local levels in the forestry sector. First, national governments collect revenues from the forestry sector and they provide forest-related services at the local level. This arrangement can be described as deconcentration of forest-related services through the national government's field offices. Second, national governments collect revenues, which are then distributed to local governments to finance forest-related services and/or basic public services in other sectors. This pattern transfers policy responsibilities to subnational governments, which remain accountable to the national government. Third, subnational governments collect revenues from forests at the local level and they are responsible for delivering forest-related services. Under this arrangement, local governments are provided with the power to decide on the utilisation of forest resources and the level of forest-related services provided at the local level. Finally, national governments collect revenues, but subnational governments are responsible for providing forest-related services at the local level. There is no direct transfer to finance the services from the national to local levels.

The fourth pattern appears to be common in decentralised forest management in developing countries. Even when the revenues are distributed back to local governments, the distribution is often not allocated directly to finance forest-related services performed by subnational governments (for example the case of revenue distribution in Indonesia). Local governments often have to use their budget to finance forest-related activities. As forest services create spillover benefits, local decision makers then neglect the benefits that spill over to other jurisdictions and only provide the services below the efficient level. This situation is obvious in the case of conservation, when the benefits of conservation reach beyond local administrative boundaries (Ring et al., 2010). Decisions on the designation of protected areas are usually made by national governments. However, conservation results in costs for local governments as they are restricted to raising revenues from taxes or charges in forests designated as conservation areas (Ring et al., 2010). To address this situation, fiscal decentralisation theory suggests that the central government should subsidise local governments to provide efficient levels of the activity on their own (Inman and Rubinfeld, 1997).

Public actors have a set of economic or fiscal instruments to regulate the behaviour of their citizens within a locality. The instruments can influence the economy through government public revenues and spending decisions (Musgrave, 1959; Ring, 2011). Many of these instruments have been used to control pollution, although they are less popular in conservation or natural resources management (Table 2.2). Government can use taxes, charges and fees to provide specific public services, although taxes are not specifically returned to those who pay the tax or earmarked for specific purposes. Government can reduce taxes, or provide subsidies and payments, for private actors who conserve biodiversity (Ring, 2011).

On the public expenditure side, national governments can consider using IFTs to correct the mismatch between revenues and expenditures and to subsidise local governments for providing efficient levels of forest services. The use of IFTs in environmental management has been explored mostly in relation to pollution control. As polluting activities in one jurisdiction flow across boundaries, the central governments can directly tax the polluting sources or provide a subsidy to local governments so that they will internalise the benefits from interjurisdictional pollution control (for example Oates, 2001; Sigman, 2005; Kunce and Shogren, 2005). Some authors have

Table 2.2 Economic instruments for public actors in pollution control, conservation and natural resources management

Instrument	Application in pollution control	Application in conservation and natural resource management
Charges/taxes	Direct charges based on quantity or quality of a pollutant	Payment for use of natural resources, including environmental services. Taxation, charges and fees to internalise positive externalities of conservation
Subsidies	Financial payments designed to reduce damaging emissions	Financial payments designed to conserve scarce resources
Marketable emission permits	Emissions reduction credits or cap-and-trade	Emissions credits from reduced degradation, such as deforestation in REDD+
Performance bonds	A deposit paid, repayable on achieving compliance in environmental quality standard	A deposit paid, repayable on achieving restoration of an ecosystem
Liability payment	Payment in compensation for damage, e.g., restoration of sites polluted by illegal dumping	Payment in compensation for damage to natural resources

Source: Modified from Perman et al. (2011, p. 182) and Ring (2011).

recently suggested the importance of assigning revenues to local governments to pursue biodiversity conservation (Ring, 2002; Ring, 2008a; Ring, 2008b; Ring, 2008c; Ring et al., 2010; Köllner et al., 2002; Kumar and Managi, 2009). IFTs in support of biodiversity conservation have been applied in Brazil and recently in Portugal. Brazil is the country with the longest standing experience with ecological fiscal transfers, which have resulted in an increase in both public and private protected areas (May et al., 2002; Ring, 2008c). An empirical study conducted by May et al. (2002) reveals that protected areas in Paraná increased by 165 per cent over a period of nine years since the beginning of the programme in 1992. IFTs have also been proposed for Switzerland, Germany and India to compensate the forgone opportunity costs by the localities with protected areas (Köllner et al., 2002; Ring, 2002; Ring, 2008a; Ring, 2008b; Kumar and Managi, 2009). In order to understand how intergovernmental fiscal transfers operate, Chapter 5 discusses the important aspects of the design of IFTs and provides examples of IFTs to support biodiversity conservation.

2.4 CONCLUSION

In most decentralised countries, national governments control revenues generated from natural resources, including forests. This situation is a common practice in developing countries where national governments are considered to have better capacity in handling complex revenue systems from natural resources and dealing with revenue distribution to subnational governments. On the other hand, local governments have a comparative advantage to perform forest-related services, such as forest protection, fire monitoring and management as well as local forest users' empowerment, due to their close proximity to forest areas and local people in surrounding forests. However, local governments are often not provided with sufficient resources to perform these services. This situation creates a mismatch between the revenues generated and the expenditures to provide forest-related services. Forest-related services also create spatial externalities because the benefits of conservation spill over across local jurisdictions' boundaries, while the costs are borne by local stakeholders. The mismatch between revenues and expenditures as well as the spatial externalities discourage local governments from pursuing good forest management and conservation.

To address the issues of the mismatch between revenues and expenditures and spatial externalities, fiscal decentralisation theory suggests assigning greater authority to raise local revenues, or increasing transfers from the central to local levels. Given the existing situation in decentralised countries in general, the latter option appears to be more feasible; thus, IFTs become a cornerstone of subnational government financing. IFTs can serve the purpose of correcting the mismatch between expenditures and revenues generated locally, as well as internalising spatial externalities related to forest management at the local level. In biodiversity conservation, some studies have analysed the different options for the distribution formula to determine the amount of the transfer to each local unit. However, the related issues of conditionality and accountability of IFTs for biodiversity conservation have received little attention. Whilst local governments should be provided with the authority to allocate public resources according to local priorities, it is also important to ensure IFTs for conservation are used to support conservation activities at the local level, hence, earmarking may be necessary. Chapter 5 will address these aspects.

NOTE

1. Upfront fees are collected on the basis of the sale of a right either to explore or exploit natural resources. Rent income is the annual revenue generated from the utilization of natural resources; while a back-end payment is the rent income received in the form of public infrastructure built by the firm as part-payment for the extracted material (Searle, 2007, p. 392).

3. The cases of Riau and Papua provinces

Indonesia provides an interesting example of the implementation of intergovernmental fiscal transfers for conservation. Following decentralisation in 2002, district governments now have the authority to provide basic public services. The Regional Autonomy Laws do not specifically regulate the forestry sector, and Santoso (2008) argued that, according to Law 32/2004, the forestry sector is not a mandatory affair or a minimum public service to be performed by local governments. This means that local governments can choose whether to partake in the forestry sector to increase the welfare of the people within their jurisdictions. According to the latest Regional Autonomy Law 23/2014, the power to manage forests is mainly distributed between the national and provincial governments. However, as we will discuss below, district governments still have an important role in decision-making processes concerning the exploitation of natural resources.

To achieve the objective of this book, two case studies – Riau and Papua provinces – were selected to inform the design of the IFT for conservation. They were chosen on the basis of the following factors: (i) local governments' commitment to, and interest in, conservation (or reducing deforestation); (ii) fiscal capacity; and (iii) special autonomy status. Riau has the highest deforestation rate in Indonesia, while Papua currently has one of the lowest rates of deforestation in the country. Riau is one of the richest provinces in Indonesia as it is one of very few oil- and gas-producing provinces. On the other hand, Papua has low fiscal capacity. Papua is highly dependent on unconditional transfers and a special autonomy fund, which is distributed only to regions that have special autonomy status such as Papua and Aceh, located at the northern tip of Sumatra Island. Furthermore, the Provincial Government of Papua enjoys a special autonomy status that provides it with full authority within all sectors of administration, except for foreign affairs, security and defence, monetary and fiscal management, religion and justice. The Provincial Government of Papua was therefore allowed to issue special regulations to manage forest resources in the province because of its special characteristics. In contrast, without a special autonomy status, the management of forest resources in Riau needs to refer to the national regulations.

In what follows, we provide the context for forest management in Indonesia and in the two cases of Riau and Papua. We first discuss the overall forest management arrangements in Indonesia by reviewing the existing regulatory framework. The account of the two provinces is provided through a thick description of the case studies, Riau and Papua. This provides the basis for the discussion in the next chapter of the perspectives of provincial and district governments, and the factors influencing their commitment to conservation and to reducing deforestation.

3.1 NATIONAL POLICIES ON FOREST MANAGEMENT

The Forestry Law 41/1999 stipulates that the state, or the Ministry of Forestry, is responsible for the management of forests in Indonesia.[1] All forested lands are categorised into state forests (*Kawasan Hutan Negara*) and private forests (*Hutan Hak*), the latter held under a private rights regime. Approximately 120 million hectares, or 62 per cent of the total land area of Indonesia, are deemed to be state forests areas (Contreras-Hermosilla and Fay, 2005), although not all of them have actual forest cover.

Specific functions are assigned to state forests, determining the purpose that a specific area should serve. The major forest zone functions comprise production, protection, and conservation. The main function of *production forests* is to generate forest commodities, mainly timber. Some production forests are also classified as *conversion forests*, which can be legally converted to other non-forest land-use activities, including infrastructure, agriculture, tree crop plantations and mining. To use conversion forests, there should be a transfer of authority from the Ministry of Forestry to the National Land Agency. *Protection forests* provide environmental services such as hydrological regulation, flood prevention, erosion control, avoidance of seawater intrusion and maintenance of soil fertility. *Conservation forests*, which include national parks and nature reserves, are intended to maintain biodiversity.

A number of production activities are currently allowed by Law 41/1999 in the exploitation of Indonesia's forests. Productive activities that can take place in production forests include the commercial extraction of timber from natural forests and plantations, extraction of non-timber forest products for commercial and household needs and commercial utilization of environmental services, such as the generation of carbon credits (Table 3.1). Following decentralisation, the subnational level governments are authorised to issue several forest utilization permits (Table 3.1). However, the authority to issue utilisation permits for commercial timber in production forests, which is a very lucrative activity, is retained by the national government. The financial importance of the activities that are under the authority of local governments is less significant compared to commercial logging. Local authorities are required, therefore, to cooperate with the central government if they wish to maximise timber revenues (Barr et al., 2006). Local governments only have the authority to provide recommendations to the Ministry on the issuance of commercial timber utilisation permits.

Since decentralisation started in Indonesia after the fall of President Suharto, three decentralization-related regulations have been issued. First, Law 22/1999 transferred the authority to provide a number of public services to the district/municipality level. Eleven important functions were devolved to the local level, including education and culture, health, agriculture, industry and trade, communications, public works, environment and land management. Law 22/1999 was revised by Law 32/2004, which introduced several changes, including: (i) the direct election of heads of district and provincial governments; (ii) greater clarity on the functions that local governments are obliged to carry out, and (iii) a reaffirmation of the role of provinces as representatives of the central government. In general, government at the provincial level has the authority to: monitor the performance of governments at the district level; play a coordination role between the district and national levels; and coordinate and monitor the implementation of deconcentrated affairs, which are the national government's responsibilities being implemented at the provincial level.

In 2014, a new Regional Autonomy Law was issued to clarify further the distribution of power between government levels in many sectors and to confirm the authority of provincial governments.

Table 3.1 Types of permits in the forestry sector

Permit type	Maximum area/ volume/duration	Activities	Issuing authority
Commercial timber utilisation permit	No maximum volume or area 55 years for natural forest 100 years for plantation forest	All activities related to the harvesting, processing, marketing, planting, and management of timber species in designated areas	Ministry of Forestry based on recommendations from heads of districts or provinces
Commercial non-timber forest product utilisation permit	No maximum volume or area 10 years for natural forest 100 years for plantation forest	All activities related to the harvesting, processing, marketing, planting, and management of non-timber forest products (NTFPs) in designated areas	Head of district/ municipality for forest areas within one district or municipality
Environmental services utilisation permit	1,000 hectares 10 years	Activities that utilise an area to provide environmental services without damaging its natural ecosystem (incl. ecotourism, water use, carbon trading and biodiversity conservation)	Governor for areas crossing district or municipality boundaries Ministry of Forestry for areas crossing provincial boundaries
Commercial forest estate utilisation permit	50 hectares 5 years	Activities that utilise a living space without disrupting the area's principal function	
Timber exploitation permit	20m^3 1 year	Activities involving timber harvesting to meet individual needs and/or public facilities of communities in forested areas	
Non-timber forest product exploitation permit	20 tonnes 1 year	Activities involving non-timber harvesting to meet individual needs and/or public facilities of communities in forested areas	

Source: Resosudarmo et al. (2006, pp. 49–50).

Each of the aforementioned regulations changed the structure of power in the management of forests between the national and local governments, as discussed below.

Soon after decentralisation, Forestry Ministerial Decree 05.1/2000 granted local governments the authority to allocate large commercial timber concessions of up to 50,000 hectares in a single district or province. District governments were also allowed to issue small-scale forest product exploitation rights and timber exploitation and utilisation permits for areas up to 100 hectares in conversion and in production forest areas (Barr et al., 2006). Following that transfer of authority, most of the forest-rich districts aggressively issued permits (Barr et al., 2006). The distribution of large numbers of small-scale timber permits may have facilitated illegal logging in some regions (Resosudarmo et al., 2006). In 2002, the Ministry of Forestry revoked the authority of district and provincial governments to issue large commercial timber concessions and small-scale logging concession permits.

The implementation of the 1999 Forestry Law was detailed by Government Regulation 06/2007, which stipulates that the management of forests in Indonesia will be carried out through Forest Management Units (FMUs). The organisations that need to be set up to manage each FMU have the following tasks and responsibilities: (i) manage the forest area, including developing a forest management plan, forest utilisation, forest zones utilisation, forest rehabilitation and reclamation, forest protection and conservation; (ii) detail the national, provincial and district policies to be implemented by local forest plans; (iii) implement the management of forests in their location, including planning, organising, implementing and monitoring and evaluation; and (iv) promote investment to support the achievement of forest management objectives. The Minister of Forestry has the authority to designate a forest zone as an FMU with a specific function, including protection, conservation or production forest. On the other hand, the authority to establish an organisation to manage the FMU is the responsibility of different governmental levels. If the forest boundaries fall under one district only, the district government will have the responsibility to establish the FMU's organisation. When the forest unit stretches across districts, the provincial government will be responsible for establishing the organisation to manage the FMU, while, if the forest area crosses the boundaries of one province, it will be under the authority of the

Ministry of Forestry. The Ministry of Forestry aimed to establish 120 FMUs by the end of 2014. To date, 120 ministerial decrees formalizing the locations of the FMUs have been issued, with around 15 of them assigned to local governments for management, with the remaining FMUs supposed to be managed by the national government through its local offices.[2]

The latest Regional Autonomy Law 23/2014 further clarified the distribution of authority over forest management in Indonesia. The only authority devolved to the district level in this respect is the management of grand forest parks (*Taman Hutan Rakyat*). The remaining management authority is distributed between the national and provincial levels of government, including:

- Forest planning: the authority to develop forest plans remains with the national government. It includes carrying out forestry inventory, formalising forest boundaries, assigning functions to forests and developing the national forest plan.
- Forest management: the national government also maintains the authority to exploit forests, particularly for commercial timber permits (i.e. commercial timber utilisation). Activities such as carrying out forest rehabilitation, reclamation and protection, as well as issuing permits for forest product processing are under the authority of the national government. Provincial governments are responsible for establishing some FMUs, as noted above, and they have the authority to issue permits for commercial forest estate utilisation, environmental services utilisation (except for carbon sequestration), and non-timber forest product exploitation.

Although local governments have limited formal authority over forest management, in practice they have almost de facto control over forest resources, as the national government has limited capacity to properly manage all areas classified as forest zones due to its geographical distance from them. Even in the case of commercial logging activity, over which the national government retains the authority to issue permits, the support of local governments is important to ensure companies' operational activities on the ground. This is because high levels of resistance from local stakeholders can hinder the implementation of licences issued by the national government, as reported in a number of regions in Indonesia.[3] Moreover, local governments are authorised by Law 26/2007 on spatial planning to develop local spatial plans according to guidelines and norms established by the national government. Spatial plans regulate the purpose of a specific zone, such as cultivation, conservation, and infrastructure development. The spatial plan approach is a hierarchical and complementary system, starting at the national level down to provincial and district/regency level. The national plan is a long-term strategic plan with a timeline of 25–50 years. Provincial plans are medium-term strategic plans over a period of 15 years and municipal and district plans are short-term operational plans with a timeframe of 5–10 years. Spatial plans can be revised every five years at all levels to adjust the function of an area in accordance with its physical condition.

Based on local spatial plans, local governments can submit proposals for land-use change to the Ministry of Forestry within lands classified as conversion forests. A change in forest status can take place in conversion forests if the proposal for land-use change meets all of the following criteria: (i) the activity has strategic importance; (ii) it does not have negative impacts on the environment (an environmental impact assessment is required to accompany the proposal); (iii) it does not lead to the creation of enclaves (removing one part of the forest for non-forest purposes) or cut the forest into a mosaic; (iv) it does not affect river flow; and (v) it has been approved by the local parliament. If the criteria are met, the change in forest status is approved by the Ministry of Forestry with a decree. In the case of forest clearance for oil palm plantations, local governments have the authority to issue a business permit, which is required before the final decision on forest clearance can be made by the Minister of Forestry (Colchester et al., 2006).

A number of productive activities are permitted by law to exploit Indonesia's forests and often cause forest degradation and deforestation. Commercial logging is normally the first activity allowed to open up natural forests legally. This activity requires the issuance of a commercial logging concession, which is granted for a 20-year period to perform selective timber-cutting based on the guidelines provided by the Ministry of Forestry (Kartodihardjo and Supriono, 2000). Over the past forty years, commercial logging operators have failed to implement sustainable forest management (ibid.). According to 1998 data, 16.57 million hectares out of 69.4 million hectares under logging concessions were degraded (ibid.). Forest degradation data for 2004 show that this trend continued into the first part of the 2000s (Nawir et al., 2007). The total area of degraded production forest amounts to 14.2 million hectares, with an additional 13.6 million hectares of logged-over areas. The area of degraded forest inside the protection forest category in 2004 was reported at 8.1 million hectares (ibid.).[4]

Despite the destruction of natural forests caused by logging operators, the government continued to issue logging permits in natural forests to generate revenues and employment (Kartodihardjo and Supriono, 2000). Licences granted to logging operators that caused severe forest degradation at the end of the concession period were terminated in some cases, and the degraded forest handed over to a state-owned company for rehabilitation (Kartodihardjo and Supriono, 2000; Nawir et al., 2007). However, one of the so-called rehabilitation policies actually involves converting severely degraded forest to commercial timber plantations. The underlying concept was to replace forest vegetation (with a remaining standing stock of less than 16 m^3 per hectare) with fast-growing species such as acacia (Kartodihardjo and Supriono, 2000; Nawir et al., 2007),[5] thus legitimising forest degradation within commercial logging areas (Kartodihardjo and Supriono, 2000).

Conversion forests are not supposed to have significant tree cover or timber potential. In reality, however, primary forests can also be found in areas designated as conversion forests. For instance, approximately 3.6 million hectares of conversion forest in Papua were assessed to be primary forest (Ministry of Forestry, 2008b). One of the major drivers of the massive conversion of Indonesia's natural forest is the establishment of oil palm plantations (Butler et al., 2009; Koh and Wilcove, 2007; Sandker et al., 2007; Venter et al., 2009). Conversion of natural forests to oil palm plantations provides additional profits for plantation companies because timber is harvested and sold during land-clearing at the beginning of operations. As a result, companies seek to acquire areas larger than the area that will actually be planted (Forest Watch Indonesia and Global Forest Watch, 2002; Kartodihardjo and Supriono, 2000).

All the aforementioned activities generate revenues for government stakeholders. In commercial logging, the national government collects revenues in the form of fees and taxes, which are then distributed amongst governmental levels using a revenue sharing mechanism. If forests are converted to crop plantation activities, the government also obtains revenues from other taxes and charges.

Other local governments' revenue sources related to plantation activities are local fees on agricultural products. Following decentralisation, local governments are required to generate revenues from their own taxes and fees as well as the shared revenues from the national government. REDD+ measures could therefore limit the revenue stream of local governments due to the restriction to pursue the aforementioned productive activities in forestlands (Gregersen et al., 2010). On the expenditure stream, local governments would be required to perform additional services to support good forest management to ensure the successful implementation of REDD+. Services, such as better monitoring and the management of conservation and protected areas, would need to be strengthened in order to ensure REDD+ achieves the targets to reduce deforestation and forest degradation (Nepstad et al., 2009). Both impacts would shift local governments' fiscal capacity to deliver public services for local residents.

Having presented the relevant features of the regulatory framework for forest governance, we now turn to the factors that influence local officials' interest in, and commitment to, conservation and reduced deforestation. The two cases chosen for this study portray different characteristics in terms of the interest of local officials. Each of the cases is first presented in detail, including the biophysical situation, fiscal capacity and forest conditions. This discussion will provide background before presenting the results of the cross-case comparison presented in the next chapter.

3.2 STORIES FROM RIAU

Riau province is one of the most natural-resource abundant regions in Indonesia. It is one of only five oil- and gas-producing provinces (World Bank, 2007b). However, revenue inequalities between districts in Riau are high. The two districts that were studied in depth for this study, Siak and Pelalawan, provide good examples of the inequality issue. Siak district enjoys considerable benefits generated from the involvement of the local government enterprise in managing an oil field in the district. In contrast, Pelalawan district derives the majority of its revenues from its share of fees and taxes generated by the forestry sector, which are trivial compared to the revenues from oil in Siak district. Unlike Pelalawan district, the Siak government no longer depends on transfers from the national government.

Generally, the topography of Riau province, and the case study districts, is lowland plains with a slope of 0–2 per cent and is about 10 metres above sea level. Siak district is suitable for agriculture and plantations.[6] Most of the land area in Pelalawan district is lowland plains with an altitude between 3 and 6 metres above sea level. The low lands are usually peat land and alluvial rivers.[7] Riau also has a good road system; about 4,200 kilometres of national and provincial roads in 2013.[8] These biophysical conditions allow easy access to most land areas, including forests.

Riau currently has the highest deforestation rate in Indonesia. Uryu et al. (2008) estimate the total loss of Riau's forests between 1982 and 2007 at around 4.2 million hectares or approximately 65 per cent of its original forest cover. The remaining forests in 2007 were reported at 2.5 million hectares. Annual deforestation between 1982 and 1988 was slightly higher than 150,000 hectares (Uryu et al., 2008). The rate dropped considerably between 2000 and 2002 to below 100,000 hectares, but then increased again. Data from the Ministry of Forestry (2008b) show that annual deforestation within the forest zones in Riau between 2003 and 2006 was about 157,688 hectares, consisting of 59,560 hectares in protected and conservation forests, 45,559 hectares in production forests and 52,569 hectares in conversion forests. This total annual deforestation is lower than the figure reported by Uryu et al. (2008) of around 198,000 hectares annually over the same period.

The major cause of deforestation in the province is the expansion of oil palm and timber plantations. Logging concessions within production forests are being replaced by acacia plantations (Figure 3.1) to supply the pulp and paper industry. Between 1982 and 2007, 1.1 million hectares (28.7 per cent) of Riau's forests were replaced by oil palm plantations, while 0.95 million hectares (24.4 per cent) were cleared for acacia plantations (Uryu et al., 2008). Uryu et al. (2008) also report that 659,200 hectares (17 per cent) of forest areas were deforested but had not yet been planted; the remaining lands were cleared for smallholder oil palm plantations (7.2 per cent) and other purposes (18.1 per cent), such as infrastructure, rubber, coconut and other plantations. These data point to the fact that Riau's forests were already in a dire condition even before decentralisation in 2001,

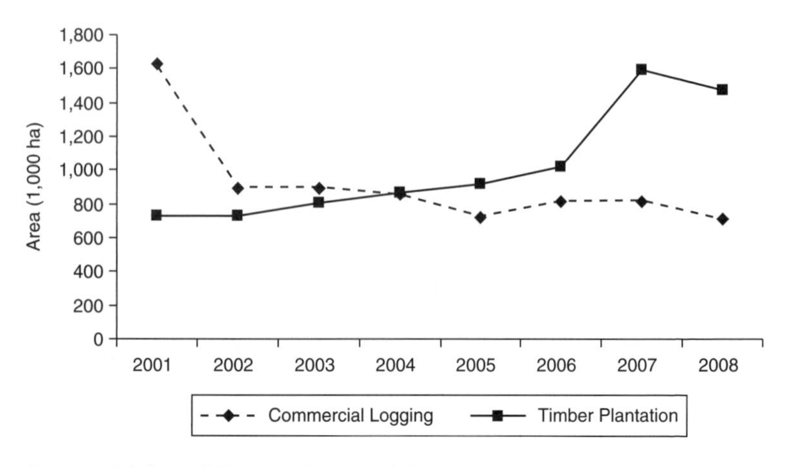

Source: Ministry of Forestry, Forest Statistics, 2008

Figure 3.1 Logging concessions and timber plantations in Riau's production forests (total area)

although greater authority provided to local government soon after decentralisation also contributed to more forest destruction (ibid.). Barr (2000, p. 6) reported that:

> Because the increases in processing capacity have far outpaced HTI (timber plantation) development, all of Indonesia's pulp producers have until now been dependent on mixed tropical hardwoods (MTH) obtained through clearing of natural forests . . . Indonesia's pulp industry has accounted for approximately 835,000 hectares of deforestation over the past 12 years. It is notable that virtually all of this area was cleared to supply wood to four large mills and that a single mill – Indah Kiat Pulp and Paper owned by Sinar Mas/APP – accounted for over one-third of the total area deforested.

As in every other forest-rich region in Indonesia, local governments in Riau took advantage of the greater authority over forest management provided to them following decentralisation. The issues reported from Riau were specifically related to the expansion of timber plantations rather than the issuance of small-scale commercial logging extraction permits as occurred in other regions of Indonesia.

> Probably the longest standing claims of illegal activities in the pulp and paper sector have focused on the allocation of small-scale conversion concessions (known as IPPK or *Izinpemungutandanpemanfaatankayu*) to RAPP (Riau Andalan Pulp and Paper) and APP (Asia Pulp and Paper) (Singer, 2009, p. 163).

Both RAPP and APP took advantage of the loophole in decentralisation legislation that existed before 2003 by expanding their areas of operation to carry on receiving timber extraction and utilisation permits, which allowed companies to clear-cut the remaining forest. Most timber plantations were established at the expense of natural forests, which resulted in a growing environmental campaign (ibid.).

In relation to oil palm plantations, the area devoted to this land-use was 40,000 hectares in 1982, and it had reached 1.5 million hectares by the end of 2006 (ibid.). Oil palm plantations provide rapid and reliable income for poor families. Due to the declining timber industry, in early 2000 the Ministry of Forestry approved the designation of certain areas within the forest estate for oil palm plantations (ibid.). The Governor of Riau, too, who was presiding over the province during the mid 2000s, was supportive of the establishment of oil palm plantations, particularly for smallholders, to combat poverty.[9] He was re-elected for a second term in 2008, following an election campaign during which he promised, among other things, to launch a poverty eradication program involving the promotion of oil palm plantations at the smallholder level by allocating an additional 50,000 hectares to poor families (ibid.).

The national government is responsible for the issue of massive licences in Riau leading to forest clearance. As of 2012, around 162 forest relinquishment permits, covering 1.75 million hectares, were issued for oil palm plantations in Riau (Riau Provincial Forestry Office, 2013). Only six licences out of the total 162 licences, covering an area of 54,673 hectares, were issued by the local government. A similar trend can also be observed in relation to commercial logging, where all licences – five concessions covering 256,788 hectares – were issued by the national government. With regard to timber plantations, the national government had issued 18 concessions (covering just over one million hectares) prior to decentralisation, and six units (of slightly less than 250,000 hectares) after decentralisation. During 2002 and 2003, the provincial government issued 34 licences covering an area of 375,477 hectares (Riau Provincial Forestry Office, 2013).

3.3 STORIES FROM PAPUA

Papua is home to the largest area of remaining tropical forest in Indonesia, and it also has the lowest rate of deforestation in the country. Forest Watch Indonesia and Global Forest Watch (2002) estimated that 1.8 million hectares were deforested between 1985 and 1997, representing 5 per cent of Papua's total forest cover. Andrianto et al. (2008) report that in the period 1997–2000, total forest loss was roughly 2.9 million hectares in Papua and West Papua.[10] However, deforestation slowed between 2001 and 2005, with total forest loss at 0.57 million hectares (ibid.). Tropenbos International (2010) reports a lower figure of total deforestation between 2000 and 2005 (430,620 hectares). The total annual deforestation within forest zones in Papua province alone between 2003 and 2006 was 19,481 hectares (The Ministry of Forestry, 2008b). The difference between deforestation reported by the Ministry of Forestry (2008b) and Andrianto et al. (2008) as well as Tropenbos International (2010) is due to the fact that the latter two studies apply to the two provinces of Papua and West Papua.

The province of Papua is one of the three provinces to have obtained special autonomy status, the others being Aceh and West Papua. According to Law 21/2001 on Special Autonomy Status, the provincial government of Papua is provided with authority within all sectors of administration, except for the five strategic areas of foreign affairs, security and defence, monetary and fiscal affairs, religion and justice. The provincial government is authorised to issue local regulations to further stipulate the implementation of the special autonomy, including regulating the authority of districts and municipalities within the province. Due to its special autonomy status, Papua province is provided with a significant amount of special autonomy funds, which can only be used to benefit its indigenous peoples. But the province has low fiscal capacity and it is highly dependent on unconditional transfers and the above mentioned special autonomy fund, which accounted for about 55 per cent of the total revenues in 2008.

After obtaining its special autonomy status, in order to allow the local population access to timber production benefits, the Papuan provincial government issued a number of decrees, enabling:

- a Timber Logging Permit for Customary Communities, which enabled local people to carry out timber extraction in small

concessions (250 to 1,000 hectares) for one year through a community-based or participatory community cooperative;[11]
- a Permit to Manage Customary Forests, which was a timber extraction permit for larger concessions (up to 2,000 hectares) for a maximum of 20 years;
- logging companies had to pay compensation to local communities in addition to all other fees and taxes collected by the national government.

After decentralisation, local governments' aggressiveness in issuing local regulations that related to timber had occurred not only in Papua. Generally, all forest-rich district governments in Indonesia took advantage of their expanded authority to issue district-level timber concessions, as noted for instance by Barr et al. (2006). However, many local regulations were in conflict with the national government's forestry laws, and district governments also showed little capacity to regulate the activities of the companies receiving the licences. This led to a situation described as 'legalising' illegal logging (ibid.) and in 2002 the Ministry of Forestry restricted local governments' authority to issue new logging and forest conversion permits in state forests.

In 2005, a report by Telapak and the Environmental Investigation Agency (respectively, an Indonesian and an international NGO) noted that the community logging rights in Papua had been subverted by illegal logging. A number of opportunistic individuals and companies took advantage of the confusion and proliferation of legal loopholes following decentralisation to set up a complex international network of timber smuggling from Papua to China to extract as much timber as possible (Singer, 2009). Many operations focused almost exclusively on Merbau (*Intsia bijuga sp.*), a type of hardwood used for top-end construction and furniture. Responding to this situation, the national government revoked the decrees issued by the provincial and district governments related to all community right licences,[12] including licences under the timber logging permit for customary communities and the permit to manage customary forests (Barr et al., 2006; Singer, 2009). Furthermore, participatory community cooperatives were dismantled.

The Indonesian President also instructed the police to conduct a 'sustainable forests operation' to combat illegal logging.[13] The operation seized timber across the whole of the Papua region on the grounds of the lack of legal documents such as transport permits (Singer, 2009), with more than 400,000 cubic metres of logs and sawn timber being confiscated, together with a host of trucks, ships and logging equipment (The Ministry of Forestry, 2005). Furthermore, 'more than 170 people were arrested, including police, army and forestry officials' (Telapak and EIA, 2006, p. 2). In May 2010, another operation was launched, but it seized a relatively smaller amount of timber, standing at 5,300 cubic metres of illegal Merbau. The operation was fuelled by information provided by the Chinese government following the arrival of illegal timber in China.[14]

In 2006, Barnabas Suebu, the first directly elected Governor of Papua, took office. Suebu made plans to declare a moratorium on log exports and recommended that no more logging concessions be granted to timber companies (Tedjasukmana, 2007).[15] In 2008, a Local Special Regulation on Sustainable Forest Management was issued. The main features of this regulation involve: (i) allowing indigenous people (local community groups) to manage forests; (ii) banning log exports from Papua; (iii) revoking commercial logging licences unless the companies have timber processing facilities and/or community logging programmes; and (iv) supporting forest conservation to generate environmental services. In 2008, the governor signed a Memorandum of Understanding with New Forests, an Australian firm specialising in environmental markets, to deliver carbon credits to the voluntary market. Governor Suebu was named a 'Hero of the Environment' in 2007 by *Time Magazine* for his role in pursuing forest conservation in the province. *Time Magazine* reported in an article about Suebu (Tedjasukmana, 2007):

> 'Why would we cut down trees if people are going to pay us to protect them?' he asks. 'We can prevent deforestation and also use the money to reforest the areas in critical condition.' Suebu says that the legal autonomy the province has when it comes to resource management will help him take on Jakarta. 'Pressure on our forests is coming from the forestry department because they are still operating with an old mindset,' he explains. 'They need to realise that there is a new paradigm now and we are not going to repeat the mistakes of the past.'

The next chapter will examine the influence of different factors in shaping local governments' perspectives and commitment to forest conservation or reducing deforestation.

NOTES

1. In 2014, the Ministry of Forestry was merged with the Ministry of Environment. The ministry is now known as the Ministry of Environment and Forestry. Since this research was carried out from 2009 to 2013, we refer to the ministry as the Ministry of Forestry throughout this book.
2. Ministry of Forestry, 2014, Regulasi dan Pembiayaan KPH, Disampaikan pada Semiloka NKB 12 K/L KPK Jakarta, 12 November 2014, Direktur Wilayah Pengelolaan dan Penyiapan Areal Pemanfaatan Kawasan Hutan, Direktorat Jenderal Planologi Kehutanan Kemen LHK. http://acch.kpk.go.id/documents/10157/1817970/Regulasi+dan+Pembiayaan+KPH.pdf
3. http://www.riaumandiri.net/rm/index.php?option5com_content&view5article&id514559:datangi-kantor-bupati-meranti-ratusan-massa-demo-tolak-hti-&catid540:riau-raya accessed on 18 November 2010. http://dpd.go.id/2010/01/dpd-desak-menhut-hentikan-izin-usaha-hutan accessed on 18 November 2010.
4. Forest degradation is defined as forested lands that are severely impacted by intensive and/or repeated disturbances, thus reducing the capacity of forests to supply goods and services (Nawir et al., 2007).
5. The precise definition of unproductive forests, however, varies. Since 1986, large-scale timber plantations must be located on unproductive forests, which are defined as having from 5 to 20 m^3 of commercial timber per hectare (Pirard, 2008).
6. http://riauprov.go.id/index.php?bahasa=&mod=subhal&sublink=potensi_daerah_siak accessed on 18 November 2010.
7. http://www.bkpmd-pelalawan.go.id/topografi.htm accessed on 18 November 2010.
8. http://riau.bps.go.id/index.php?hal=publikasi_detil&id=1414 accessed on 6 February 2015.
9. Based on Local Regulation 2/2006.
10. Soon after decentralisation in 2003, Papua was divided into two provinces: Papua and West Papua provinces. Data published on Papua's forests, even after the separation, still mostly include both provinces.
11. Governor Decree 522.2/3386/SET 22 August 2002.
12. Ministerial Decree P.07/Menhut-II/2005.
13. Presidential Instruction No. 4/2005.
14. http://www.walhi.or.id/en/categoryblog/1545-moratorium-konversi-hutan-untungkan-makelar accessed on 18 November 2010.
15. In August 2014, Suebu was announced as a suspect in a corruption case related to a hydropower project in Papua. In April 2015, the Corruption Eradication Committee again announced Suebu as a suspect of another corruption case, which is also related to the previous case. Mr Suebu was still in detention in mid 2015.

4. Factors affecting local forest governance

In Chapter 2, we noted that the decentralised forest management literature points to a number of factors that influence local governments' decisions about forest management. Several empirical studies (Andersson et al., 2006; Andersson et al., 2004; Andersson, 2004; Andersson, 2003; Larson, 2002; Larson, 2003; Ribot et al., 2006) have considered the influence of some of those factors – such as sufficient resources and capacity, financial incentives, upward accountability, discretionary power, demands from NGOs and pressure from local people – on *successful* decentralised forest management. For the most part, those studies do not examine deforestation (or conservation) as the dependent variable. They use *good forest management*, which is generally defined as the interest of local governments in undertaking forest initiatives and providing forest-related services. Although forest services provided by local governments, such as forest rehabilitation, may contribute to reducing emissions from the forest sector, reducing deforestation is more complex than simply delivering forest-related services at the local level. Two studies that specifically examine factors influencing local governments' interest in avoiding deforestation find a number of important attributes, including: NGO pressure, local financial importance of forestry, socioeconomic context, national policy, local institutional performance and biophysical factors (i.e. access to forests and topography) (Andersson and Gibson, 2007; Andersson et al., 2010). These studies appear to increase the emphasis on the financial and economic aspects of forest resources compared with other related studies.

Deforestation can be perceived as either good or bad by local stakeholders – including local communities and government officials – since legal land-use changes and forest exploitation generate revenues. Based on the perspectives of the local officials, this chapter therefore seeks to understand how financial incentives as well as other factors influence local governments' interest in reducing deforestation and pursuing conservation. It is also important to consider factors other than financial incentives because the latter alone may not be sufficient to shift local governments' perspectives and interests.

4.1 METHODS

A causal mechanism approach is adopted to analyse qualitative data collected through interviews with local governments. To explain the complex view of a social situation, a case study approach with small-N can apply a mechanism approach to causation to understand how causes interact in the context of a particular case, or a few cases, to produce an outcome (Bennett and Elman, 2006). Mahoney (2001) contends that causal mechanism analysis is different from correlational analysis. Correlational analysis identifies 'antecedents regularly conjoined with outcomes'; consequently, the increase or decrease of a causal variable has the probability of generating higher or lower values on an outcome (Mahoney, 2001, p. 580). In contrast, a causal mechanism consists of identifying 'the process that underlies and generates empirical regularities and outcomes', where the activation of a mechanism is sufficient to produce the outcome of interest (ibid.). Causal mechanisms are useful to explain why correlations exist in the first place and/or to suggest new correlations that have not been previously discovered (ibid.).

Causal mechanism analysis is common in political science scholarship that seeks to explain a complex social world (Bennett and Elman, 2006). A causal mechanism can, however, be considered speculative, particularly when relatively little is known about the causal relationship between variables, and for this reason, research on causal mechanisms is best governed by more established theories (Gerring, 2010). Mahoney (2000), for instance, suggests that explanatory variables should not be randomly selected for consideration. Rather, they should be chosen on the basis of the theoretical literature relevant to the research questions. Therefore, the established theories of decentralised forest management were considered in an earlier chapter in order to select the explanatory variables and to reduce the uncertainty of the proposed causal mechanism.

The two techniques applied in this study to uncover a causal mechanism are process tracing and nominal comparison. First we identified potential factors that may influence the interest of local officials in reducing deforestation and pursuing conservation using the within-case (process tracing) analysis. The technique of difference was then applied to eliminate causes that are not sufficient to influence the interest of local officials in conservation (or deforestation). With the technique of difference, possible factors, with similar values, that are found in both cases can be eliminated because they are not sufficient causes to generate the outcome. Each technique is detailed below.

Primary data on the factors that shape local governments' interest in reducing deforestation and pursuing conservation were gathered through in-depth interviews. The interview data were further verified through local policy documents and regulations as well as the actual deforestation rates in the case study provinces. In-depth interviews were carried out with 25 individuals, who were selected through purposeful sampling. All respondents from several government agencies (Finance Department, Forestry Department and Development Planning Agency) at provincial and district levels held formal decision-making power. The interviews were carried out in 2009 (Riau) and 2010 (Papua), and again in 2013 (in both provinces) to ascertain possible changes in perspectives. Some officials had moved on to new positions within government agencies. The interviews were carried out on the basis of the position in the government structure rather than the particular person. Hence, in some locations, new officials were interviewed due to changes in staff. The perceptions of local government officials can be influenced by their personal values and experience. However, these officials are the ones who implement government activities related to forest management or development in general. Therefore, their perspectives matter.

4.1.1 Within-case Analysis: Process Tracing

Causal process tracing is used to identify and verify within-case implications of causal mechanisms. Process tracing is 'a technique to locate the causal mechanism by linking a hypothesised explanatory variable to an outcome' (Mahoney, 2000, p. 409). Process tracing is an operational procedure for identifying and verifying the observable within-case implications of causal mechanisms through examining traces for every step between the cause and the outcome (Blatter and Blume, 2008). Causal process tracing requires a full storyline with density and depth of events within their contexts. In order to provide the explanations, it is important to describe the finest level of events and realities that are observed (Blatter and Blume, 2008).

In small-N case study research, within-case analysis is a tool specifically designed to compensate for limitations associated with cross-case methods (Mahoney, 2000). This tool is useful to avoid mistaking a spurious correlation for a causal association, which arises when 'two correlated variables appear to be causally related but in fact are the product of an antecedent variable' (ibid., p. 412). Causation is not established through small-N comparison alone, but through 'uncovering traces of a hypothesised causal mechanism within the context of an historical case or cases' (Bennett and Elman, 2006, p. 262). The within-case analysis attempts to specify the beginning and ending of the temporal context in which the causal process plays out, because social processes do not occur instantly (Falleti and Lynch, 2009), and this obviously applies to an issue such as the commitment to conservation and deforestation reduction.

In each of our cases, Riau and Papua provinces, this study attempts to identify 'temporal unfolding situations, actions and events, traces of motivations, evidence of interactions between causal factors, and detailed features of a specific outcome' (Blatter and Blume, 2008, p. 319). To achieve this objective, information was gathered not only through direct interviews with stakeholders but also from published materials such as local regulations, reports, academic papers, magazines and online databases. The published materials can complement the qualitative data collected through direct interviews to understand the perceptions and motivations of the actors related to deforestation and forest conservation. The rich qualitative data are then analysed to identify the causal structures behind the existing situation of forest management in each locality.

4.1.2 Cross-case Analysis: Nominal Comparison

Following the within-case analysis, a cross-case comparison was conducted by applying nominal comparison analysis. Nominal comparison involves the use of categories that are mutually exclusive (cases cannot be classified in terms of more than one category) and collectively exhaustive (one of the categories applies to each case) (Mahoney, 2000). Nominal comparison involves giving vivid labels in the format of nominal categories. Unlike ordinal comparison – which 'entails the rank ordering of cases into three or more categories based on the degree to which a given phenomenon is present' (ibid., p. 399) – no specific rank or degree is assigned to the nominal categories. A causal argument, or different values of the dependent variables, are summarised into categories, where each of the cases is classified either as a member or non-member of the categories.

Although often considered unsophisticated, as it does not involve the ranking or ordering of cases, nominal comparison is highly appropriate to conceptualise certain kinds of phenomena (ibid.). It can be used to identify sufficient or necessary causes of an outcome (ibid., p. 392).

> When a sufficient cause is present, the outcome will always also be present. However, if a sufficient cause is absent, the outcome could be either present or absent. If a necessary cause is absent, the outcome will always be absent. However, if a necessary cause is present, the outcome could be either present or absent.

To eliminate potential necessary causes, the method of agreement can be used, whilst the sufficient causes can be eliminated using the method of difference (Mahoney, 1999, 2000). The method of agreement requires that the outcome of interest is present in all cases, thus, any hypothesised cause that is not shared by all cases cannot be considered to be a necessary cause. In contrast, with the method of difference, in some cases the outcome is present while in others the outcome is absent, so a hypothesised cause that is found in all cases cannot be a sufficient cause (Mahoney, 1999, 2000).

Mahoney (1999, 2000) provides an example of nominal comparison used by Skocpol (1979) in identifying the causes of state and social revolutions. Potential causes were first divided into four explanatory variables: (i) conditions for state breakdown; (ii) conditions for peasant revolt; (iii) relative deprivation; and (iv) urban worker revolt. She further assessed whether or not each of the causes existed in her study cases (France, Russia 1917, China, England, Russia 1905, Germany, Prussia and Japan). Mahoney (1999) explains how Skocpol (1979) eliminated the potential causal factor of social revolution. Using the method of difference, the factor of relative deprivation was eliminated as it appears in both positive and negative cases of revolution. The urban worker revolts factor was eliminated using the method of agreement, since the factor was not present in all three cases (France, Russia 1917, and China), where a social revolution occurred.

4.2 RIAU: IS DEFORESTATION BAD?

4.2.1 Local Governments' Perceptions on Deforestation and Conservation

Interviews with local government officials in Riau reveal factors that influence local governments' interest in, and commitment to, reducing deforestation and pursuing conservation. Responses provided by local government officials can be categorised into four factors: (i) beliefs and values towards forest resources, (ii) pressure from local people, (iii) authority devolved, and (iv) financial incentives. These four factors were consistently mentioned throughout interviews with district and provincial officials in Riau. Each of these factors is discussed below.

In 2009, interviews with local officials revealed that they did not consider deforestation to be a problem. According to Riau's forest classification, almost half of the forestland is classified as conversion forest. Hence, land-use change in this forest category is legal. In areas classified as production forest, logging activities are also permitted by law, although local officials also pointed out that a number of silvicultural practices contribute to forest degradation. Moreover, local officials perceived deforestation as a change of forest classification (from forest zone to non-forest zone). Thus, the conversion of natural (or logged-over) forests to timber plantations (monoculture timber species) was not considered to be deforestation since timber plantations were still classified as production forest. This perspective had changed by 2013 when local officials thought that the era was now one of 'conservation'. As demands for forest conservation increased, local officials believed that national and local policies should also shift from merely extracting forest resources to protecting them.

Local officials believe that land-use change and forest exploitation are required to increase people's welfare and to pursue regional economic development. The belief and values of local officials can be linked to the pressure of local people, most of whom seek lands to pursue their livelihoods. Local officials also think that the needs of local people are not similar to those of outsiders (environmental activists or NGOs), who want to see more forest conservation. In the end, local officials in Riau admitted that they are more responsive to the pressure from local people than to the environmental activists' demands for more forest protection.

> [Conservation] is their [NGOs'] need, then they try to change us. Reports about deforestation and forest degradation in Riau are being exaggerated . . . According to Greenpeace and Jikalahari [local environmental NGO], forests in Riau are currently in a dire state however local people think the opposite (Riau Provincial Forest Agency Interviewee #1, 2009).

In general, the relationship between the government and NGOs improved between 2009 and 2013. During interviews in 2013, local officials mentioned that they had started working together on conservation efforts with several environmental NGOs.

Respondents consistently noted that local communities needed to be allowed to pursue their livelihoods in forests even if their activities result in forest destruction. If the national government or outsiders are interested in conservation, sufficient financial resources need to be provided to compensate what local people are currently gaining from forest conversion. As stated by an official from Riau Provincial Forestry Agency, 'if now we are told to stop cutting trees, people should be compensated'.

Interviews with provincial and district officials revealed discontent related to the power devolved under the existing decentralisation setting. The frustration regarding the authority devolved was revealed during interviews in 2009; unfortunately, the situation remained the same in 2013, where local officials provided more recent data to support their arguments. Local governments think that basically no authority is provided to the local level in terms of forest management and they currently only act as the implementers of national policies. Local officials also believe that limited financial resources constrain the implementation of national policies in the forestry sector at the local level.

When resources are transferred to local governments, they are usually inflexible in terms of how the funds can be spent. Activities to be implemented at the local level vary depending on the local situation, an aspect which is often neglected during policy formulation at the national level. District officials also mentioned that district governments often have to bear the costs of conservation at the local level. The costs not only include the forgone benefits from production forests that are being allocated to conservation, but also from supporting conservation management at the local level.

4.2.2 Within-case Analysis

Further tracing of the motivations that lay behind local officials' interest in supporting land-use change revealed four other important factors. First, local officials perceive deforestation as necessary for economic development and people's livelihoods. Second, all respondents noted that local governments must take into account local people's needs in developing policies. Local people require land to grow oil palm for their livelihoods, thus, the conversion of forest is considered important for local people in Riau. Since forest conservation always comes with a restriction on the pursuit of productive activities in forests, local officials argued for the importance of financial resources to compensate local people for the livelihoods that would be restricted by forest conservation. Third, related to forest services to be provided by local governments, many national policies cannot be implemented at the local level due to limited financial resources and authority. Fourth, limited authority and power devolved to local governments is considered a constraint to pursuing good forest management, because even if local officials were committed to conservation or reducing deforestation, they would not have the power and resources to do so, unless they obtained the blessing from their national counterparts. This argument was supported by the fact that licences to deforest are mostly given by the national government. Only a small portion of licences was issued by local governments after decentralization in 2002–2003, as noted in the previous chapter.

To assess factors that shape local governments' interest in, and commitment to, forest management in Riau, the interview data were triangulated with major events that are relevant to forest management in the province. A number of factors seem to play a role in shaping local governments' interest in, and commitment to, forest management in Riau (Figure 4.1). Several events that can be observed through the literature and media records, as discussed in Chapter 3, include: (i) Presidential Instruction 4/2005 on combating illegal logging; (ii) numerous NGO protests; (iii) prosecution of corruption cases in the forestry sector; and (iv) discussions on REDD+ that started in 2010. Two events that appear to be

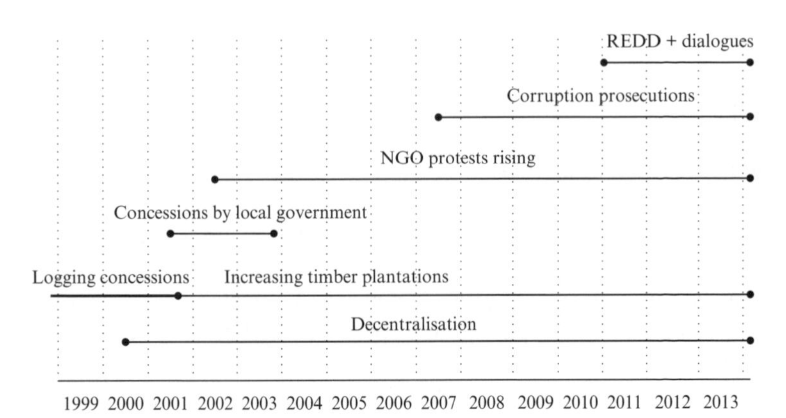

Figure 4.1 Timeline and context of forest management in Riau

influential in forest management are: (i) NGOs' protests, and (ii) corruption prosecution as a form of law enforcement (coercion) conducted by the national government. Local officials were reluctant to talk about these issues – particularly about corruption cases – as some investigations were still being carried out during the interview phase of this study in 2009. During interviews in 2013, Riau was about to elect a new government and, in 2014, the newly elected Governor of Riau was detained by the Corruption Eradication Committee for a corruption case on forest relinquishment for developing an oil palm plantation. Although these two factors are not mentioned as important factors by local officials, they shape the perspectives of local officials on forest management, particularly when a violation of law related to land-use activities is investigated or prosecuted.

A shift in local officials' perspectives could be observed between 2009 and 2013. First, there was a change in the relationship between NGOs and the government. In 2009, NGOs seemed to position themselves as an adversary of the government, while in 2013, local officials mentioned that they had started working together with local NGOs in several forest-related activities, including REDD+. Second, the discussions about REDD+ had changed the perspectives of local officials about the way they perceived conservation. In 2013, local officials were less defensive about deforestation and mentioned that the 'forest exploitation' era had ended and the 'conservation' era had started. Finally, there was a change in the government of Riau province and several other districts.

Whether the apparent shift in the perspectives of the local government in Riau to forest protection influenced deforestation may be assessed by considering the findings of remote sensing studies. The Ministry of Forestry (2012) reports that there was a decline in the deforestation rate from around 152,861 hectares annually in 2006–2009 to only 74,167 hectares per year in 2009–2011. However, Margono et al. (2014) reported that there was an increase of deforestation in Riau from around 206,675 hectares annually in 2006–2009 to 280,047 hectares annually in 2009–2011. The latter source was peer reviewed, therefore it should be considered more reliable. Therefore, it would appear that the reported change in local government perspectives did not influence the deforestation rate. Whether the effect of changes in local government perspectives have lagged effects would need to be assessed.

4.3 PAPUA: COMMITMENT TO CONSERVATION

4.3.1 Local Governments' Perceptions of Deforestation and Conservation

In April 2013, Governor Lukas Enembe took office from Barnabas Suebu. Interviews carried out in 2013 revealed that the perspectives of local officials remained consistent between early 2010 and the end of 2013. A local official mentioned that the new governor was similarly committed to sustainability. Governor Enembe is seen as more independent, less influenced by international NGOs when compared with Mr Suebu. Local officials revealed that forest conservation had become the top priority of the governor. The provincial government intends to set aside 60–70 per cent of the total land area in Papua Province for conservation and protection purposes. Conservation and protection of forests, as designated by the Ministry of Forestry, currently account for 45 per cent of the total forestland in the province. Although it is important to recognise that some forests are not easily accessible, this target is significantly higher than the national minimum standard of all localities in Indonesia, which is 30 per cent of the total land area. Governor Enembe has also launched *Papua 2100*, a sustainable spatial plan that aims to provide guidance for land management and monitor its implementation. Provincial officials perceive that conservation has a positive impact on their localities. They are aware of the need to prevent forest destruction by learning from the mistakes of their counterparts in other parts of Indonesia. Their beliefs and values towards conservation and forest protection came out strongly in every interview. At the district level, the commitment to forest conservation is associated with delivering forestry services such as forest rehabilitation, planting and reforestation. However, officials at this level do not seem to completely understand the idea of avoiding deforestation, as they are more project-oriented.

The pressure of Papuan communities related to deforestation varies between districts. Local officials in Jayapura District revealed that they currently must deal with forest encroachment conducted by local people who migrated from other districts. The migrants are already causing destruction in one of the nature reserves in the district. Upton (2009) reported that Jayapura District has the lowest numbers of non-indigenous residents. At the provincial level, the pressure is generally considered low because most of the indigenous Papuans reside in the central highland area, which is located in the middle of Papua province (BPS Papua, 2009). Many indigenous people have had limited exposure to the outside world, a factor that could be contributing to a reduced pressure to deforest.

> Some of our people are still living in *jaman batu* [literally translated as 'stone-age'] . . . If they are asked to maintain their forests, they are very motivated (Papua Provincial Planning Agency Interviewee, 2010).

Local officials expressed less frustration about the distribution of power compared with their counterparts in Riau district. Due to Papua's special autonomy status, they seem to have more bargaining power with the national government in terms of policy implementation. However, district officials noted some issues related to the management of conservation areas, expressing their willingness to become more involved in conservation activities, which are currently managed directly by the national government.

Local officials in Papua did not express specific concerns about the availability of financial resources for forest management; however, they pointed out that those resources are important to ensure the quality of services provided by subnational governments. Despite their interest in protecting and conserving forests, local governments cannot implement actions without sufficient financial resources. For instance, local officials argued that forest conservation had not been optimal due to the low fiscal capacity of local governments. Most of the local units in forest management funded by the local budget are only equipped with a motorcycle and a limited number of staff. Furthermore, an official from the Provincial Planning Agency stressed that, with financial assistance provided by developed countries, Papua could reduce carbon emissions generated, including in the forestry sector:

> Now, we are waiting for Australia, England and America whether they want to assist us achieving our targets to reduce emissions. The Governor is very committed and we have already conducted socialisation with the local people (Papua Provincial Planning Agency Interviewee, 2010).

At the time of writing (early 2015), financial support for reducing deforestation in Papua Province had yet to materialize.

4.3.2 Within-case Analysis

The perspectives of local officials noted above were triangulated with the major events related to forest management in Papua, which were discussed in the previous chapter. Key events were (Figure 4.2): (i) Mr. Suebu's leadership and the special autonomy status of local governments, which influenced local governments' interest in, and commitment to, pursuing forest conservation at the local level; (ii) following the issuance of Presidential Instruction 4/2005, the 'sustainable forest operation' which also took place in Papua in 2004–2005 and 2010; (iii) the issuance of local regulations related to forest management in the province. To take advantage of the special autonomy status, the Provincial Government of Papua issued Special Provincial Regulation 21/2008 on forest management, and the Governor Regulation 13/2010 was also issued to allow indigenous communities to manage forests.

Interviews with local officials in Papua highlighted factors that influence their interest in conservation similar to those in Riau, but they appear to have a stronger focus on conservation. Local officials' perspectives appear to be influenced by the following issues: (i) they have sufficient authority to make decisions related to forest

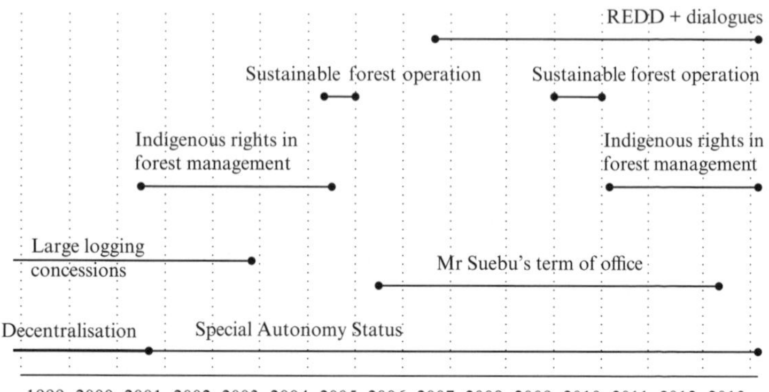

Figure 4.2 Timeline and context of forest management in Papua

management due to their special autonomy status; (ii) there is low pressure from local people, as a considerable area of the land is not easily accessible for productive activities, and local indigenous people live in forests, with basic and non-destructive living styles; and (iii) there are potential financial incentives for supporting conservation.

The situation at the district level is similar to that at the provincial level, but there are some small differences. In several areas, for instance in the Cycloop Nature Reserve in Jayapura District, local people are trying to encroach upon the forest. Jayapura and Sarmi districts are mostly located in the lowland regions with easy access to forest areas. These districts are also home to the majority of migrants living in Papua province. Moreover, strong leadership and commitment at the provincial level have not completely translated at the district levels, where local officials are mostly concerned about performing forest services such as rehabilitation, planting and community development.

The special autonomy status of the province also enables local officials to bargain with the national government to prioritise local policies. The provincial government can take decisions according to its priorities and negotiate with the national government whenever there are contradicting policies. At the district level, local officials generally do not appear to have major concerns about the authority devolved for forest management. They did, however, mention problems with the management of conservation areas, which come under the authority of the national government. They noted that improved coordination is required between the national governments' staff located at the local level and district officials. Finally, local officials in Papua noted that lack of financial resources has hindered the implementation of forest management at the local level. Local governments at provincial and district levels in Papua have also anticipated financial assistance that might be provided by bilateral and multilateral donors to support conservation in their localities. This anticipation appears to play a role in influencing local governments' perspectives towards conservation.

Despite the limited financial resources, local policies and efforts in forest protection and conservation seem to have resulted in the reduction of deforestation in Papua. According to data from the Ministry of Forestry, deforestation decreased between the period of 2001–2009 (12,744 hectares annually) and 2009–2011 (7,157 hectares annually). A similar trend was also reported by Margono et al. (2014), although the precise rate of deforestation is not available.

4.4 CROSS-CASE ANALYSIS

Having analysed the responses provided by local government officials in Riau and Papua, at least seven factors influencing their interests and commitment to forest conservation and reducing deforestation were identified. Five of the seven factors were explicitly mentioned by local officials: (i) local officials' beliefs and values, (ii) pressure from local people, (iii) devolved authority, (iv) financial incentives, and (v) leadership. Two other factors were not specifically mentioned but were apparent in both provinces: (i) the prosecution of corruption, and (ii) NGO protests and campaigns.

In order to select factors that have sufficient influence on local governments' interest in, and commitment to, deforestation or conservation, a nominal comparison between cases was conducted (Table 4.1). The interest and commitment of local governments were assessed from the responses provided by local officials during interviews and were also verified from the implementation of local policies and the enactment of local regulations on forest conservation. For instance, local officials' interest and commitment to conservation in Papua were further demonstrated by the issuance of local regulations on sustainable forest management as discussed above.

Table 4.1 Factors influencing local governments' interest in conservation

	Papua 2009	Papua 2013	Riau 2009	Riau 2013
Deforestation rate	Declining		Possibly increasing	
Interest in, and commitment to, forest conservation	High	High	Low	Low to medium
Beliefs and values towards conservation	High	High	Low	Low to medium
Pressure from local people	Low	Low	High	High
Authority devolved	High	High	Low	Low
Leadership	Present	Present	Not present	Not present
Relationship with NGOs	Friendly	Friendly	Not friendly	Friendlier
Corruption prosecution	Yes	Yes	Yes	Yes
Anticipated financial incentives for conservation	Yes	Yes	No	Yes
Actual flow of funds	No	No	No	No

Local governments in Riau and Papua appear to have opposite interests and commitments to reducing deforestation and pursuing conservation (the dependent variable). Local officials in Papua province, starting from the Governor, perceive reducing deforestation and promoting forest conservation as important priorities for their localities. In contrast, the Governor of Riau perceived the need to support oil palm plantations for local people's livelihoods,[1] with local officials believing that forests are important resources to sustain those livelihoods. At the provincial level, local officials are candid about the fact that the deforestation that is occurring in Riau is legal and therefore is not a problem. Officials stressed that local people, rather than NGO protests, shape their beliefs and values about forest management, as their main responsibility is to improve people's welfare. This was also noted by Singer (2009), who reports that the pressures from local people are significant in Riau, and sometimes lead to conflicts that threatened the safety of local officials. As the position of NGOs changed and became more cooperative with local governments, local officials showed an interest to work together with local NGOs in conservation. Hence, they appeared, at least in words, to share the same value.

In Riau, officials at provincial and district levels think that there is basically no power devolved to the local level in terms of forest management. They think that it is their role to cater for local people's voices and needs in forest management, however, they have no authority except to implement the national government's instructions. This arrangement puts them in a difficult situation. In contrast, the Papuan provincial government has the advantage of greater authority in forest management with a stronger bargaining position with respect to the national government. It can put pressure on the national government to take local situations into account in the decision-making process.

With the method of difference, factors (with similar values) that are found in both cases can be eliminated because they are not sufficient causes to generate the outcome (Table 4.1). One factor with a similar value that is found both in Riau and Papua is corruption prosecution. This factor can possibly be eliminated as a sufficient cause to shape the interests and commitment of local officials towards forest conservation.

In Riau, there was a change in the beliefs and values of local officials towards forest conservation and deforestation between 2009 and 2013. First, the relationship between NGOs and the local government changed in Riau between 2009 and 2013 as NGOs are more recently considered friendlier as they work together with local governments. Another change that occurred in Riau between 2009 and 2013 was the onset of an intense dialogue about REDD+ in the province. Hence, the good relationship between NGOs and government, and the anticipation of financial incentives due to REDD+ dialogues, could result in changing the beliefs and values of local officials. However, that change would not necessarily result in real actions that lead to reducing deforestation and promoting forest protection.

Local officials in Riau and Papua noted that financial incentives are important to enable them to reduce deforestation and pursue conservation. Since 2007, the provincial and district governments in Papua have been anticipating financial assistance that might be provided by bilateral and multilateral donors to support conservation. In an interview conducted in 2007, the Governor of Papua had already proclaimed his interest in pursuing forest protection because it may generate income from carbon credits (Tedjasukmana, 2007). In contrast, the discussion about the potential support from international donors to preserve forests in Riau started in 2010. During the interview in 2009, local officials were not convinced about the feasibility of REDD+, or financial support for conservation. However, there was a shift in perspective about REDD+ following the signing of the $1 billion agreement between Indonesia and Norway. This dialogue started at the national level and was later followed by the provincial dialogue to develop a REDD+ strategy.

The impact of REDD+ dialogues may have changed the interest and commitment of local governments towards forest conservation. However, at the time of writing, there had been no financial flow to local governments. To translate the interest and commitment into actions, financial resources for financing forest conservation are crucial, particularly when local governments have limited budgets.

4.5 THE CAUSAL MECHANISM

Following the within- and cross-case analyses presented above, we turn to considering a possible causal mechanism of local governments' interest in, and commitment to, forest conservation and reducing deforestation (Figure 4.3). While the within-case analysis has identified the causal factors that may influence the interest of local governments in pursuing conservation and reducing deforestation, the cross-case analysis has helped eliminate factors that are not sufficient to influence that interest.

Local governments' beliefs and values contribute to their commitment towards forest conservation and reducing deforestation. These beliefs and values are heavily influenced by the pressure of local people. Local governments, particularly district governments, perceive their role as the frontline of bureaucracy that has to address the needs of local people. Deforestation and forest conservation can therefore be perceived as either good or bad. The provincial government of Papua sees a need to conserve more forests and reduce deforestation. This commitment is then put into action by allocating more

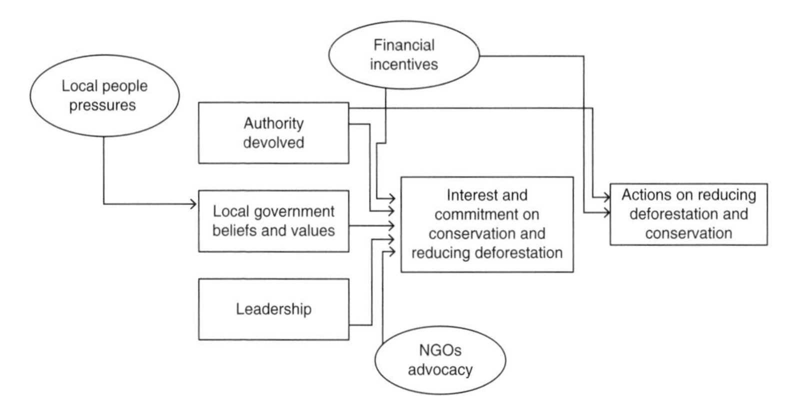

Figure 4.3 The causal mechanism of local governments' interest and commitment to reducing deforestation and pursuing conservation

forestland to conservation, which is beyond the national minimum standard, and issuing necessary regulations to better manage forest resources. However, when deforestation is considered necessary, and good for people and regional development, deforestation and forest degradation can be expected to continue. Furthermore, the case of Riau shows that improved relationships between the local governments and NGOs contributed to the change of the commitment and interests of local officials in conservation and reducing deforestation.

The variables included in the proposed causal mechanism are not exhaustive, and cannot be generalized to other case studies. This study does not exclude other variables that may influence the beliefs and values of local governments towards conservation. Beliefs and values are complex mental events. They 'entail thought, which can be differentiated according to their ascending levels of generality e.g. specific programmes, issues-area doctrines or policy paradigms, ideologies or public philosophies and cultures' (Yee, 1996, p. 69). Sabatier (1988, 1991) argues that the policy-making process, particularly related to agenda setting, is dominated by elite opinion; as a result, the impact of public opinion is at best modest. On the other hand, Kingdon (1995) suggests that the process of defining issues and agendas is shaped by public opinion. Specifically in forest management, the commitment to the long-term sustainable use and conservation of forest resources can be influenced by local governments' 'social and environmental ideology' (Larson, 2003) and the personal characteristics of the heads of local governments, including their education (Andersson et al., 2006).

Local governments' beliefs and values would, however, have less impact if they have limited or no authority in the decision-making process. The devolved power appears to determine the interests and commitment of local governments in pursuing conservation and reducing deforestation. Transfer of meaningful power and sufficient resources to autonomous local authorities are considered imperative for the implementation of decentralised forest management (Ribot et al., 2006; Ribot, 2003). However, to maintain control, central governments in developing countries often devolve obligations rather than meaningful powers (Ribot et al., 2006).

Leadership of district heads or governors influences the interest and commitment of local officials to conservation. However, research on the impact of leadership on the interest and commitment of local governments towards conservation and/or deforestation is limited. Nagendra (2007) analysed the relationship between leadership and forest change in Nepal and found an insignificant association between the two variables. However, the author explained that the results may have been influenced by the difficulty in measuring leadership and that the effectiveness of leadership could vary over time. Moreover, NGO protests and coercive actions by national governments, such as corruption prosecution, have been described as effective corrective actions to prevent wrongdoing and punish legal violations (Larson, 2002; Andersson, 2003; Andersson et al., 2006; Gibson and Lehoucq, 2003). These factors need to be considered, although our analysis shows that they could be eliminated as a result of nominal comparison.

Finally, financial incentives appear to play a significant role in influencing local governments' interest in, and commitment to, conservation. REDD+ funds could be distributed to influence local governments' beliefs and values, particularly for those who do not see deforestation as problematic, as in the case of Riau. Sabatier (1988) suggests that elites' belief systems, particularly related to decisions in the policy implementation stage, could be altered through a process known as policy orientated learning: a process involving alteration of thoughts, or behavioural intentions, that results from experience (Sabatier, 1988). Local governments could be expected to change their beliefs and values as they experience positive responses from local people related to REDD+ implementation. Furthermore, sufficient financial resources are important for implementing actions in conservation and reducing deforestation.

The factors influencing local governments' interest in, and commitment to, conservation (and reducing deforestation) identified in this chapter are similar to the important determinants of successful decentralised forest management suggested by Larson (2003) and Tacconi (2007). Authority and resources devolved to the local level are similar to the 'legal structure' and 'capacity' mentioned by Larson (2003). Pressure from local people was highlighted by Tacconi (2007) as 'livelihoods' in his model. Local governments' values and beliefs, however, have received less attention in the literature on decentralised forest management, although Larson (2003) briefly mentions the importance of a long-term commitment towards sustainable development. Leadership, which is emphasised here, was not mentioned by those two earlier models.

4.6 CONCLUSION

The causal models presented in the literature on successful decentralised forest management may be applied as a tool to develop policies and activities for reducing deforestation and forest degradation at the local level, but adjustments may be required. This chapter confirms that financial incentives play an important role in influencing local governments' interest in, and commitment to, conservation and reducing deforestation. Financial incentives are required to finance forest services at the local level, but also to compensate for livelihood and revenue losses at the local level – the latter being used to provide services, including health and education. Financial incentives need to be present, together with a number of other factors, such as sufficient devolved authority in forest management, local leadership in conservation, and incentives for local communities, to favour conservation over land-use change. Local governments need to be provided with the decision-making power over forest resources under their administrative boundaries and they need to be allowed to pursue local priorities according to the demands of their electorates. More authority does not always lead to more conservation (or reduced deforestation) because of the influence of the values and beliefs held by local governments and the pressure of local people.

However, a policy that is developed without considering the perceptions of the different stakeholders is likely doomed to fail. Reducing demands from local people for land-use change can be brought about by creating incentives for local communities to prefer conservation to land-use change. Finally, coercive measures should also be pursued when violations occur.

In order to design intergovernmental fiscal transfers (IFTs) to distribute REDD+ revenues to the local levels, it is therefore important to consider the factors that need to exist, together with the financial incentives, to reduce deforestation. We now turn to government stakeholders' perspectives on the design of IFTs, taking into consideration the existing administrative and political situation in Indonesia.

NOTE

1. However, this perspective could be questioned on the basis that, in some cases, elected officials appeared to support land-clearing for personal gain, as they have been accused of corruption as already noted.

5. Intergovernmental fiscal transfers and Indonesia's experience

Intergovernmental fiscal transfers (IFTs) have become a cornerstone of subnational government financing. Transfers from central to local levels can address both the mismatch between revenues and expenditures and spatial externalities. To ensure IFTs achieve their policy objectives, attention has to be devoted to their design and the conditions in which they operate. The main elements of the design of IFTs are the distribution formula, conditionality and accountability (Bird, 1999, p. 24). The distribution formula entails the distributable pool (grant size), which is the vertical dimension of IFTs that determines the total amount of grants or transfers distributed to different government levels, and the horizontal dimension, which is the basis for distributing transfers to eligible local units. The literature on IFTs for biodiversity conservation suggests that political processes as well as community lobbying will influence the final design of the IFTs, although different options could be developed on the basis of scientific justification (Ring, 2008b; Köllner et al., 2002). Understanding stakeholders' perspectives is therefore important, particularly because the decision-making process related to financing mechanisms needs to involve a multiplicity of actors, who may have different values and little consensus about the problem's definition and solutions.

We first lay out the main elements of the design of IFTs, including specific examples of IFTs implemented, or proposed, to support biodiversity conservation. We then address the current arrangements and experience with IFTs in Indonesia, before exploring the purposes of IFTs for conservation based on the perspectives of government stakeholders. The perspectives of government officials were researched through in-depth interviews with 36 officials in government agencies involved in the development of IFTs, including the Finance Department, the Forestry Department, and the Planning Bureau at national, provincial and district levels. The respondents were encouraged to discuss their preferences related to an IFT mechanism to distribute REDD+ revenues to the local level.

5.1 THE DESIGN OF INTERGOVERNMENTAL FISCAL TRANSFERS

5.1.1 The Distributable Pool and the Distribution Formula

IFTs are commonly used in decentralised countries to distribute public revenues from national to decentralised levels of government. As previously mentioned, the main purposes of IFTs are to distribute a share of the national government's revenues to subnational governments (vertical revenue-sharing) and to correct for spatial externalities generated from the provision of public services (Bird and Smart, 2002; Shah, 2006). Vertical revenue sharing aims to address the mismatch between expenditure needs with the public revenues generated at the local level. Since the tax base of local governments tends to be narrow, and non-tax revenues such as user charges, rents, royalties, and fees are also limited, revenue sharing is then an option to close the fiscal gap (de Mello, 2000). Spatial externalities create an inefficient outcome of public service provision, as local decision-makers often neglect the benefits accrued to outsiders beyond local boundaries in the decision-making process (Bird, 1999). Pigou (1932, cited in Oates, 1972, p. 66) proposed that to address such spatial externalities: 'the economic unit generating the spillover should receive a unit subsidy equal to the value at the margin of the spillover benefits it creates.' By providing a unit of subsidy equal to the value at the margin of the spillover benefits, local governments are expected to provide the right amount of a public service (Oates, 1972).

Related to the magnitude of the transfers, two key aspects of the design of IFTs are the size of the 'distributable pool' and the basis for distributing transfers to each eligible local government (Bird, 1999, p. 24). The three approaches to determine the size of the transfer pool are based on (Bird, 1999; Bahl, 2000; Bahl and Wallace, 2007):

1. a defined percentage of the national government's revenues;
2. an *ad hoc* approach based on a discretionary decision that may vary from year to year;

3. a cost reimbursement based on the costs of public service delivery at the local level, defined by the national government.

Decentralised countries usually use the defined (percentage) approach or the shared-tax approach to vertically distribute revenues that are collected from taxes and fees between government levels (Bird, 1999; Bahl, 2000; Bahl and Wallace, 2007). The ad hoc approach is similar to any other budgetary government expenditures, where the parliament or president decides on an allocation to the subnational government on a discretionary basis (Bird, 1999; Bahl, 2000; Bahl and Wallace, 2007). Finally, the cost reimbursement approach decides the size of a grant pool based on a proportion of specific local expenditures to be reimbursed by the central government. Central governments usually define a service for which they guarantee to cover the costs incurred by local governments in delivering the service (Bird, 1999; Bahl, 2000; Bahl and Wallace, 2007).

Approaches to the determination of the size of IFTs to eligible local units (horizontal dimension) are based on (Bahl, 1999, 2000):

1. the origin of the collection of the tax (derivation approach);
2. a formula-based approach;
3. a total or partial reimbursement of costs;
4. an ad hoc approach.

The derivation approach determines the size of transfers to local governments based on a share of a national tax, and each local government receives an amount based on the total tax collected within their geographic boundaries (Bahl, 1999; 2000). The formula-based approach applies objective and quantitative criteria to determine the size of the IFTs among eligible local government units (Bahl, 1999, 2000). The cost-reimbursement approach distributes grants on the basis of reimbursement of costs of specified services (Bahl, 1999, 2000). There is a fine line between determining the size of the IFT using the formula-based approach and the cost-reimbursement approach (Bahl, 2000). Both approaches might use an exact equation for the distribution of the IFTs, however, the cost-reimbursement approach stipulates the level of services to be provided by the local governments. Finally, the ad hoc approach determines the magnitude of the transfer to the local unit on the basis of the decision-makers' judgment, usually without specific criteria or formulae (Bahl, 1999, 2000).

Table 5.1 Types of intergovernmental fiscal transfers

Method of allocating divisible pool among eligible unit (horizontal dimension)	Method of determining the total distributable pool (grant size)		
	Specified share of national or state government tax	Ad hoc decision	Reimbursement of approved expenditures
Origin of collection of tax	A	n/a	n/a
Formula	B	E	n/a
Cost reimbursement approach	C	F	H
Ad hoc	D	G	n/a

Note: n/a: not applicable.

Source: Bahl (1999, 2000).

A taxonomy of IFTs that considers both the distributable pool (grant size) and the distribution formula has been developed (Bahl and Linn, 1992; Bahl, 1999, 2000) (Table 5.1). Type A IFTs use a shared tax approach (derivation approach), where subnational governments are allowed to keep a share of the taxes or fees collected within their administrative boundaries (Bahl, 1999, 2000; Bird, 1999). The objective of this scheme is to ensure the stability of revenue sources for local governments and also to provide some degree of flexibility on how the funds can be spent at the local level (Bahl and Wallace, 2007; Bahl, 2000). Type B IFTs distribute a portion of the national tax among local governments on the basis of a formula. For instance, in Indonesia and the Philippines, 26 and 40 per cent of national revenue collections respectively are distributed to local governments based on population, land area and other indicators (Bahl, 2000; Kaiser et al., 2006; Fadliya and Mcleod, 2010). Type C IFTs distribute a share of the national tax to cover the costs of providing specific services at the local level, such as the costs of teacher salaries (Bahl, 1999, 2000). Type D IFTs occur where the grant is a share of national taxes that is distributed to eligible local governments based on ad hoc decisions. They are seldom mentioned in the literature.

Types E, F and G IFTs distribute grants amongst local government units objectively with specific criteria, although the distributable pool is decided on the basis of ad hoc decisions and without any objective criteria. A Type G IFT is a grant where central governments make all the decisions about who will receive the grant and how much is given to each recipient, without particular criteria. The drawbacks of the ad hoc approach include:

1. lower transparency as it is subject to political manipulation;
2. a lack of certainty that can affect fiscal planning and effective budgeting;
3. no clear link between revenue sources and expenditure responsibilities (Bird, 1999; Bahl, 1999, 2000).

On the other hand, this approach provides maximum flexibility for the central government to decide the size of the transfers distributed to local governments each year, particularly in times of budget constraints (Bahl, 1999). The ad hoc approach is usually acceptable for the allocation of funds to regions facing special needs, such as during emergencies or natural disasters (Bird, 1999; Bahl and Wallace, 2007; Bahl, 2000).

The cost-reimbursement approach (Type H) establishes the amount of the transfer based on a proportion of specific local expenditures to be reimbursed by the central government (Bahl, 2000). The cost-reimbursement approach applies quantitative criteria to decide the amount of the IFT based on local expenditures to provide public services. The central government usually defines a public service for which it will guarantee to cover the costs incurred by local government in delivering it (Bird, 1999; Bahl and Wallace, 2007; Bahl, 2000). This approach is normally used to:

1. correct spatial externalities associated with public service provision at the local level;
2. provide direct investment to high priority national needs, otherwise some local governments would underspend on services with regional and national benefits;
3. ensure uniform service provision standards across the country. However, this approach may compromise local choice and can hold back fiscal decentralisation. It also involves higher implementation costs because the central government must monitor

the compliance of local governments with national standards (Bahl, 2000).

5.1.2 International Experience with the Distribution Formula for Conservation

In the case of biodiversity conservation, type B IFTs are commonly used to distribute public resources to support conservation at the local level. In Brazil, a portion of the ICMS tax,[1] which represents the largest source of state revenues, is distributed back to local governments on the basis of ecological indicators (May et al., 2002; Grieg-Gran, 2000; Ring, 2008c). The grant size of the ICMS tax is the total Value Added Tax (VAT) collected in each state, where 75 per cent of this state-level VAT is redistributed using the derivation approach, while the remaining 25 per cent is distributed on a formula-based approach using several indicators, including land area, population and ecological indicators (May et al., 2002; Grieg-Gran, 2000; Ring, 2008c). In Portugal, the grant size is determined on the basis of a certain percentage of the General Municipal Fund, which amounts to 50 per cent of the Financial Equilibrium Fund (Santos et al., 2012). The grant size is then divided on a formula-based approach using indicators such as land area (weighted by elevation levels) and protected area, which is according to the share of protected areas in a municipality (Santos et al., 2012) (Table 5.2).

Studies advocating the use of IFTs for biodiversity conservation have mostly focused on the horizontal dimension that determines the size of the IFTs for each eligible local government area. The formula-based approach is commonly suggested to derive the size of the funds to be distributed to eligible localities (Köllner et al., 2002; Kumar and Managi, 2009; Ring, 2008b). Different indicators are used to determine the amount of the IFT allocated for each eligible locality (Table 5.2). The indicators are used as proxies for the local governments' opportunity and management costs of biodiversity conservation. Hence, one could argue that the formula-based approach, used to determine the amount of the IFTs for conservation, is rather similar to the cost-reimbursement approach. As previously mentioned, both the formula-based and the cost-reimbursement approaches are similar. The main difference is that the cost-reimbursement approach stipulates the specific services to be provided by the local governments (Bahl, 1999, 2000).

Table 5.2 International experience with IFTs for biodiversity conservation

Country (introduction of scheme)	Model	Conservation indicator
Brazil (since 1992) Source: Grieg-Gran (2000); May et al. (2002); Ring (2008c)	ICMS Ecológico (state level) Size grant: a certain percentage of state value added tax revenues Distribution formula: formula-based – lump sum transfers from state to local governments based on ecological indicators	Conservation units: – based on designated protected areas in relation to municipal area – weighted by management category of protected area accounting for relative land-use restrictions
Portugal (since 2007) Source: Santos et al. (2012)	Local financing law Size grant: a certain percentage of the General Municipal Fund (FGM), being equal to 50% of the Financial Equilibrium Fund (the latter is made up of 25.3% of the average revenues from personal income tax, corporate profit tax and value added tax) Distribution formula: formula-based – lump sum transfers from national to local governments based on ecological indicators	Natura 2000 and other protected areas In municipalities with: (i) less than 70% of protected areas (PAs) in relation to municipal area, 5% of FGM are allocated for PAs; (ii) more than 70% of PAs, 10% of FGM are distributed for PAs
Switzerland (proposed) Source: Köllner et al. (2002)	Size grant: previous compensation sum for structurally weak cantons and a part of the petroleum tax Distribution formula: formula-based – integrating biodiversity indicators to the existing fiscal transfer	Biodiversity index: – based on the cantonal plant species diversity – weighted by the degree of biodiversity integration from low to high scenario

Table 5.2 (continued)

Country (introduction of scheme)	Model	Conservation indicator
Germany (proposed) Source: Ring (2008b)	Size grant: depending on model selected Distribution formula: formula-based – integrating conservation units into general lump sum transfers (considering local fiscal need and capacity) distributing a specified amount through unconditional special transfers based on conservation units	Conservation units: – based on designated protected areas in relation to municipal area – weighted by management category of protected area accounting for relative land-use restrictions
India (proposed) Source: Kumar and Managi (2009)	Size grant: (assumed at) 1,000 billion rupees Distribution formula: formula based – integrating environmental services in the fiscal transfers that reflects conservation efforts and stock of natural resources	Forest cover Geographical area

5.1.3 Conditionality and Accountability of IFTs

In terms of conditionality, IFTs can be classified as either conditional or unconditional grants (also referred to as transfers). An unconditional grant, also termed a general-purpose grant, is provided without any restrictions on how the recipient can spend the funds (King, 1984; Shah, 2006). A lump sum unconditional grant is when the amount of the grant is fixed; while an effort-related grant is when the amount of the grant depends on the recipient's revenue effort – that is, the percentage of the recipient's revenues from taxes and charges (King, 1984). In contrast, a conditional grant, also termed a specific grant, requires the funds to be spent on activities stipulated by the grantor (King, 1984; Shah, 2006). Conditional grants can be subdivided into lump sum conditional grants and matching conditional grants (King, 1984; Shah, 2006). A lump sum conditional grant is where the recipient receives a fixed amount to be spent on a designated service; whereas a matching conditional grant is where the amount transferred depends on how much the recipient spends of its own revenues on the service concerned. Matching conditional grants require grant recipients to use their own resources to finance a specific percentage of expenditures (Shah, 2006). Two types of matching conditional grants are: (i) open matching conditional grants, where the grantor matches the level of resources with those provided by the recipient; and (ii) closed matching conditional grants, where the grantor matches recipient funds only up to a pre-specified limit (King, 1984; Shah, 2006).

Conditional grants usually aim to create incentives for local governments to implement specific programs or activities (Shah, 2006). Conditional grants require the recipients to provide a particular good or service at a specified level corresponding to the fiscal transfer. Conditional grants specify the type of expenditures that can be financed (input-based or expenditure conditionality) or focus on the achievement of the policy objectives in the service delivery (output-based or performance conditionality) (Shah, 2006; Bird, 2001). Matching conditional grants are expected to reduce the price of producing a certain public good so that the consumption level of the good can be at a socially desired level, while lump sum conditional grants have no impact on the effective price per unit of public good provision (Oates, 1972; Shah, 2006). Conditional grants do not provide any freedom for local governments about how to spend the funds, as they usually establish tight conditions.

In contrast, unconditional grants usually aim to provide general budget support while ensuring local autonomy by allowing the recipients to manage funds without restrictions (Bird, 1999). Unconditional grants are distributed with the primary objective of ensuring that all regions have adequate resources to provide goods and services at minimum standards. The funds, which can be utilised according to the recipient's priorities, usually carry no condition except regular financial auditing (Bird, 2001). Unconditional grants only have an income effect, which means that the output of the public good or service will increase because the grant will enrich the community and not because of the change in the price of providing the good or service. Subnational governments receiving an unconditional grant are free to select the combination of goods and services according to their preferences under the existing prices structure. The grant does not change the choice between public goods, but subnational governments, on behalf of the local residents, can direct their income to purchase other goods (Oates, 1972).

In conservation activities, debate on whether the transfer mechanism should be conditional or unconditional remains contentious (Ring, 2008c). To ensure that funding is spent on the management of conservation areas, it is important to prescribe certain goods and services that can be financed by the transfer. In contrast, a lump sum transfer presents the advantage of providing more financial autonomy to local jurisdictions, enabling local decision makers to allocate the funds where they are most needed.

Brazil's experience in the implementation of the ICMS-Ecológico indicates that the components of conditionality and quality evaluation criteria need to be included when developing a design for REDD+ revenue distribution. The existing mechanism in most Brazilian states does not pay sufficient attention to the quality of protected areas (Ring, 2008c). Without giving much attention to the quality of protected areas, there is a tendency to create new protected areas and/or register existing ones, but not necessarily to keeping them fully protected.

5.2 IFTs UNDER FISCAL DECENTRALISATION IN INDONESIA

The Fiscal Balancing Law 33/2004 provides local governments with the authority to generate their own revenues. Following decentralisation, they quickly exercised this authority mostly by enacting local taxes and fees. District own-source revenues remain low, whilst the provincial level has a wider tax basis that allows it to have greater fiscal autonomy (Table 5.3). Taxes collected at the provincial level include: taxes on vehicles, fuels, and ground water; while taxes that can be collected at the district level include: advertising, hotels, restaurants and sand mining. To enable district governments to perform their administrative functions, the Fiscal Balancing Law has also introduced several IFTs to balance the revenues and expenditures assigned at the local level. The IFTs account for a major portion of

Table 5.3 Subnational governments' public revenue sources in 2012

	Province		District	
	Amount (IDR trillion)	Share (%)	Amount (IDR trillion)	Share (%)
Own-source revenues	75.07	46	22.64	7
Unconditional grant	27	16.6	196.87	61.2
Shared revenues from taxes and natural resources	26.29	16.1	44.59	13.8
Conditional grants	1.3	0.8	21.51	6.7
Other revenues	32.99	20.5	36.06	11.3
Total revenues	162.65	100	321.67	100

Source: Ministry of Finance (2012).

the regions' revenues (Ministry of Finance, 2008a). The three IFTs currently implemented in Indonesia are: an unconditional grant (UG; known in Indonesian as *Dana AlokasiUmum – DAU*), conditional grants (CG; known in Indonesian as *Dana AlokasiKhusus – DAK*), and shared revenues from taxes and natural resources (SR; known in Indonesian as *Dana BagiHasil – DBH*), which is an unconditional transfer that is distributed to the local governments on the basis of commodities produced, or resources extracted, in a local government area. The importance of each revenue source varies widely. At the provincial level, the IFTs financed almost 51 per cent of the total budget in 2012, while the remaining portion was financed by local revenue including taxes and fees. The portion of IFTs at the district level is even higher. In 2012, IFTs made up around 87.1 per cent of the total budget of district governments in Indonesia (Ministry of Finance, 2013).

The objective of the unconditional grant (UG) is to address the imbalance between local expenditures and revenues. The funds (grant size) for UG are drawn from net national revenues, on the basis of a fixed share (currently 26 per cent). Districts receive 90 per cent of the allocated funds, while the remaining 10 per cent is distributed across provinces. The distribution formula for the UG is intended to cover civil servant wages of subnational governments (basic allocation) including base salaries, family assistance and other allowances, with the remaining portion distributed using the following formula (World Bank, 2007b; Kaiser et al., 2006; Fadliya and McLeod, 2010):

$$UG = BA + (FN - FC)$$

Where,

BA is base allocation (personnel spending)

FN is fiscal needs

FC is fiscal capacity.

The remaining amount (FN − FC) is referred to as the fiscal gap. Fiscal capacity (FC) is defined as the sum of a local government's own-source revenues and its entitlement to revenue sharing from taxes and natural resources. Fiscal need (FN) is calculated on the basis of the average level of spending of all local governments, which are then weighted on five indices: population, area, cost index, the level of 'human development' and the level of per capita Gross Regional Domestic Product (GRDP). Spending is defined as the sum of spending on personnel, goods and services and capital goods (Fadliya and McLeod, 2010).

The shared revenues (SR) transfers aim to distribute taxes and natural resources (fees and royalties) from the national government to the producing local governments. Taxes that are returned to the producing regions using the revenue-sharing mechanism include: personal income taxes, land and building taxes, and the transfer fees from land and buildings. The shared revenues from oil and gas, mining, fisheries, forestry and geothermal energy are known as revenue sharing from natural resources. The Fiscal Balancing Law also updated the percentage of shared-revenues from taxes and natural resources distributed between governmental levels (Table 5.4). Shared revenues from taxes and natural resources are the second largest transfers to the subnational levels. Since decentralisation in 2001, the distribution of the revenues from natural resources was mostly concentrated in a few oil-producing regions (The World Bank, 2007b).

In the forestry sector, the distribution of the reforestation levy to the district level, known as the reforestation fund, is used to finance forest rehabilitation. The levy is collected from logging extraction in natural forests, where as much as 40 per cent of the total levy collected is distributed to the producing districts, while the remaining 60 per cent is retained by the Ministry of Forestry at the national

Table 5.4 Percentage allocations for shared-revenue from taxes and natural resources

Revenue source	Central	Provincial	Producing districts	Other districts in the same province	All districts in Indonesia
Mining – land-rent	20	16	64	0	0
Mining – royalty	20	16	32	32	0
Land and building tax	9	16.2	64.8	0	10
Land/building transfer fee	0	16	64	0	20
Personal income tax	80	8	12	0	0
Forest licence fee	20	16	64	0	0
Forest resource rent	20	16	32	32	0
Reforestation levy	60	0	40	0	0
Oil	84.5	3.1	6.2	6.2	0
Gas	69.5	6.1	12.2	12.2	0
Geothermal – central share deposit	20	16	32	32	0
Geothermal – land rent and production	20	16	32	32	0

Source: Law 33/2004 and Government Regulation 55/2005.

level. The latter portion is used to finance GERHAN.[2] Before 2004, the reforestation fund distributed to the district governments was treated as a conditional grant. Currently, the fund is transferred together with other components of the shared-revenues from the forestry sector, including the forest licence fee and the forest resource rent (Subarudi and Dwiprabowo, 2007). While the funds from the licence fee and forest resources rent are unconditional in nature, the reforestation fund distributed to the local level is specifically earmarked for forest rehabilitation with stringent guidelines issued by the national government.

The contribution of conditional grants (CGs) to local governments' revenues is modest compared to other transfers (Table 5.3). According to Law 33/2004, the CGs aim to finance activities that are determined by the central government, based on national priorities to be implemented at the local level. Several sectors are financed by the CGs, including education, health, fisheries, agriculture, environment and infrastructure (including roads, irrigation and water supply). Law 33/2004 regulates the allocation of CGs on the basis of three criteria:

1. General criterion, which gives priority to districts with fiscal capacity lower than the national average. The definition of fiscal capacity in the general criteria of CGs is similar to the distribution formula of the UG.
2. Specific criterion, which considers specific regional characteristics to choose the eligible regions, including: all districts in Papua province and poor regions, coastal areas, international bordering regions, regions that are vulnerable to disasters and regions with food insecurity.
3. Technical criterion, which integrates technical considerations developed by a sectoral ministry. For instance, the technical criterion for environment-related funding is the combination of the length of rivers, the level of river pollution, total land area, and population. The CG for Environment should be spent on activities related to water quality monitoring, water pollution prevention, water resources protection and the development of environmental quality information.

The allocation of CG funds involves two stages: (i) determining regions that are eligible to receive the funds, using the criteria previously mentioned; and (ii) calculating the amount of the funds allocated to each district. The distribution formula for CGs varies significantly between sectors, as the formula is decided by the national government together with the parliament every year, based on proposals put forward by the technical ministries.

Besides the three common intergovernmental fiscal transfers discussed above, there are two further transfer mechanisms known as the special autonomy fund and the adjustment fund. The special autonomy fund is allocated specifically to the provinces of Aceh, Papua and West Papua. The adjustment fund is used for the implementation of specific national policies implemented by provincial and district governments. The adjustment fund includes: specific allocations for teachers; support for schools' operations; support to build local infrastructure; and a regional incentive fund (known in Indonesian as Dana Insentif Daerah – DID). Funding through the latter fund is provided to provinces and districts on the basis of their performance in financial management, education, economic development and welfare services.

Other than those financed through the adjustment funds, the national government finances national priorities implemented by provinces through a deconcentration fund. While the adjustment funds are transferred to the budget of local governments, the deconcentrated fund is managed by the national government with the activities implemented by the provincial government. Provincial governments submit proposals to obtain funds to the national level ministries that carry out the process of planning for the deconcentration fund. The deconcentrated activities that the national government assigns to district governments for implementation (known in Indonesian as *Tugas Pembantuan*, literally translated as or Assisting Task), are funded through another fund, the Assisting Task Fund (known in Indonesian as *Dana Tugas Pembantuan*). The proposals for specific activities are submitted by district governments, while the planning and budgeting process is carried out at the national level.

5.3 GOVERNMENT OFFICIALS' PERSPECTIVES ON THE DESIGN OF IFTs FOR CONSERVATION

After the assessment of the existing regulatory framework on intergovernmental fiscal transfers, this section discusses the perspectives of government officials about the different options for IFTs. IFTs should be designed to ensure efficient service delivery and need to be modelled according to country-specific conditions (Bird and Smart 2002; Bird 1999). Based on the perspectives of government stakeholders, this section explores three aspects of the design, namely: (i) the distribution formula (the basis for distributing transfers), (ii) conditionality of IFTs; and (iii) the accountability mechanism of IFTs to distribute REDD+ revenues. During the interviews with the 36 respondents, not all of them were comfortable providing their views on this question. Some, who perceived the question as too technical, refused to answer. Some respondents also provided more than one preferred mechanism as discussed below.

5.3.1 The Preferred Design of an IFT Mechanism

Nine out of 36 respondents considered that a revenue-sharing mechanism would be preferable to distribute revenues generated from payment for environmental services (Table 5.5). Their rationale was that a revenue-sharing mechanism entails clearer guidelines on the collection and distribution of the funds compared to other mechanisms:

> The government can obtain benefits from taxes. Once the taxes are collected, they should be immediately distributed to local governments. For instance, 50 per cent from the tax collected is distributed directly to local governments. The remaining 50 per cent, for example, can be used for surrounding communities (National Development Planning Ministry Interviewee, 2009).

The existing revenue-sharing mechanism usually transfers funds to local governments unconditionally. However, respondents, who preferred the revenue-sharing mechanism, had different perspectives on the conditionality of the transfer, which is discussed in the next section.

Four respondents favoured a conditional grant to support forest conservation to local governments. These respondents were mostly from finance agencies at the national and district levels, whilst finance officials at the provincial level preferred revenue sharing to a conditional grant. The finance agencies at the national and district levels are usually responsible for the timely disbursement of funds, and are thus more concerned with the management (budgeting and accounting) of the funds and less with the technical performance of service delivery. The preference for the adoption of a conditional grant is due to its ease in administration and clear instructions on how the funds should be spent:

Table 5.5 Preferred IFT mechanism for REDD+ revenue distribution

Mechanism	Supporting institution
Revenue-sharing (unconditional transfer)	1. Papua Provincial Finance Agency 2. Riau Provincial Finance Agency 3. Papua Provincial Forestry Agency 4. Siak Forestry Agency 5. Sarmi Forestry Agency (#2) 6. The Ministry of National Development Planning 7. The Ministry of Forestry (#1) 8. The Ministry of Finance (#2) 9. Jayapura Development Planning Agency
Conditional grant	1. Jayapura Finance Agency 2. Pelalawan Finance Agency 3. The Ministry of Finance (#1) 4. The Ministry of Finance (#3)
Other mechanism	1. The Ministry of Forestry (#2) – trust fund/grant 2. The Ministry of Finance (#1) – trust fund/grant 3. Riau Provincial Forestry Agency (#2) – deconcentration fund 4. Jayapura Forestry Agency – GERHAN 5. Sarmi Forestry Agency (#1) – funds managed by the provincial government 6. Pelalawan Provincial Forestry Agency – funds managed by local state-owned company 7. Riau Provincial Forestry Agency (#3) – channelled directly to local communities

> Currently the best transfer mechanism is a conditional grant in terms of the spending of the fund. There is also a sanction when problems are found with spending, where the next transfer can be suspended (Jayapura Finance Agency Interviewee, 2010).

Six other mechanisms were also suggested to distribute conservation finance to the local level. These were: (i) trust fund/grants; (ii) a deconcentration fund; (iii) GERHAN; (iv) funds managed by provincial governments; (v) funds administered by a local government enterprise; and (vi) funds directly channelled to local communities. While the first three mechanisms currently exist in the country and are regulated by law, the last three mechanisms do not exist and the respondents did not provide detailed descriptions of their suggested mechanisms. Hence, discussions below focus on the three existing mechanisms: grants/trust funds, a deconcentration fund and GERHAN.

Grants from a foreign source could be distributed from the national to local budgets. According to Government Regulation 02/2006, foreign grants can be transferred to local governments to:

- implement local administrative functions, basic service provision and capacity-building of the local bureaucracy; and/or
- conserve natural resources, the environment, and culture; and/or
- support research and technology; and/or
- provide humanitarian assistance.

The use of grants needs to be accounted for in the local budgeting document and activities should be reported in the budget implementation document.

Deconcentration funding is usually distributed to support functions deconcentrated to provincial governments (as the representative of the national government at the local level) or to district governments. The functions are strictly determined by the national government and this funding provides no flexibility for local governments. Activities funded are usually the national government's functions that are carried out by local government.

> The role of provincial governments in the implementation of REDD+ would be to ensure coordination, control and stimulate districts to manage forest properly. We could therefore use deconcentration funds, however the funds should reach the district level (Riau Forest Agency Interviewee #2, 2009).

GERHAN is an example of a deconcentration programme that is implemented by the Ministry of Forestry. Local governments are responsible as an implementer of GERHAN at the local level to perform activities such as: determining sites within their areas to be rehabilitated, making seeds and seedlings available, providing technical information and conducting evaluation and monitoring. These activities are financed directly by the national budget, hence there is no direct transfer to the local budget.

5.3.2 Distribution Formula for IFTs

In this section, we discuss respondents' views about the distribution formula for IFTs for conservation. A quantitative analysis of the distribution formula for the case study provinces is provided in Chapter 8.

The distribution formula was seen as a very technical issue by some respondents and only a few of them were willing to address this topic. Respondents thought that payment for environmental services should be distributed to compensate at least for the revenues from alternative land-use activities, such as agriculture, that would be forgone by subnational governments. Two officials from the Ministry of Forestry argued that it is also important to consider the downstream economic activities of forest exploitation and land-use change in the estimation of the opportunity costs of conservation:

> It is important to know not only the opportunity costs but also the forward linkages to employment, the supply of raw materials . . . for example oil palm has impacts on energy, cosmetics and fertiliser production (Forestry Ministry Interviewee #2, 2009).

Two local officials thought that local governments' share of REDD+ revenues should be higher than the amount they currently receive from alternative land-use activities. For instance, a forest official from Papua Province argued that the provincial level should also obtain a share of the revenues from the reforestation fund, which is currently only distributed between the national and district governments. Moreover, a local official in Siak District thought that districts should obtain a higher portion of the reforestation fund since many activities related to forest management should be implemented at the local level. In addition, respondents argued for a specific share of REDD+ revenues to be allocated to local people and communities residing around and in the forests. Under the existing revenue distribution arrangements in the forestry sector, there is no specific allocation for local communities from productive land-use activities.

A national official from the Ministry of Finance also suggested that the distribution of revenues could be based on the services performed by local governments to support REDD+. The distribution formula could be based on REDD+ related activities to be implemented at the local level, rather than on the amount of emission reductions achieved by the localities.

Table 5.6 Preferences about the conditionality of IFTs

Conditionality	Supporting institution
Bloc grant to specific sector but no specific condition	Jayapura Local Planning Bureau Papua Provincial Forestry Agency Pelalawan Forestry Agency Sarmi Forestry Agency Siak Forestry Agency The Ministry of Forestry (#2) The Ministry of National Development Planning
Specific purpose grant or lump sum specific grant (assigned to a specific sector with specified activities to be financed)	The Ministry of Finance (#1) The Ministry of Finance (#3) Pelalawan Finance Agency Jayapura Finance Agency Jayapura Forestry Agency (GERHAN) Riau Provincial Finance Agency – channelled to local community Riau Provincial Forestry Agency (#3) – channelled to local community Riau Provincial Forestry Agency (#2) (deconcentration fund)
No condition or earmarking (not assigned to specific sector or activity)	The Ministry of Finance (#2) Papua Provincial Finance Agency

5.3.3 Conditionality of IFTs

Only two respondents thought that conservation finance should not be earmarked (Table 5.6). Two different rationales were provided. First, earmarking would constrain local governments in pursuing local priorities. Second, earmarking could cause a mismatch between revenues and expenditures in the forestry sector at the local level. The expenditures of the forestry agency are often higher than the revenues generated from the forestry sector (Papua Provincial Finance Interviewee, 2010). The district government of Jayapura, for example, often has to subsidise public service provision in the forestry sector because of the imbalance between revenues and expenditures.

The remaining respondents thought that earmarking is preferable, although there were different perspectives about whether funds should be earmarked to a specific sector or to finance particular activities. Funds for conservation could be allocated to a specific sector, for instance the forestry sector, however, local governments should have the autonomy to propose and decide activities. This arrangement could prevent the national government from issuing too stringent guidelines that cannot be implemented at the local level.

> Funds could be allocated to the forestry sector, but we shouldn't be told that we have to build this or that or finance this or that, this approach simply cannot be implemented. (Jayapura Local Planning Bureau Interviewee, 2010)

This type of conditionality currently does not exist in Indonesia as conditional grants are usually earmarked to finance specific activities. Local agencies cannot use funds for activities other than those prescribed by the national government. When local officials want to use funds for other activities that are considered necessary for their regions, they need to propose to the national government a revision or alteration of the items to be financed. However, this process is usually long and cumbersome.

There is no particular difference in the responses provided by the national and local officials in relation to conditionality of IFTs (Table 5.6). Rather, perceptions varied across institutions. The finance agency would prefer strict conditionality. Earmarking funds for particular activities are considered on the basis of the existing capacity of local governments.

> A conditional grant is easier to be implemented and the disbursement is not complicated as we only implement activities instructed by the national government. The purpose of the funds is clear. Currently we are terrified about funding that carries some flexibility because we are afraid of making mistakes. This also relates to regions' low capacity in financial management (Pelalawan Finance Agency Interviewee, 2010).

On the other hand, the technical ministry, in this context the forestry agency, would prefer funding to be assigned to a particular sector, but local agencies should have flexibility to decide activities implemented at the local level. Earmarking to a specific sector is considered preferable, as it would allow local governments to decide activities that are most needed in their localities. Two local forestry officials thought

GERHAN and the deconcentration fund could also be used to finance REDD+ related activities at the local level. These two mechanisms are strictly conditional in nature, and funds are not transferred to the local budget. They made this suggestion because the deconcentration fund (including GERHAN) could still be used at the local level, whilst the reforestation fund currently cannot be spent in their regions.

5.3.4 The Capacity to Manage IFTs at the Local Level

The responses provided by interviewees on their preferred mechanisms also consider the existing capacity of local governments in managing public resources transferred to the local level. For instance, due to low capacity in financial management, conditional grants are preferred as they are transferred with clear guidance from the national government. This helps ensure that any mistake in spending within the fund can be prevented. It is interesting, therefore, to gain an understanding of the existing capacity of local governments in terms of financial management as background for the administrative and institutional setting of public financial management in Indonesia. The capacity of local governments in managing public resources at the local level can be assessed from the management of the reforestation fund distributed to the local level. The current circumstances encountered by local governments, which impede the use of the reforestation fund, and their perspectives of local governments on the issues are considered below.

In Riau, representatives of the Forestry agency reported that the reforestation fund had not been spent and was now accumulating in the local account. Approximately 50 trillion rupiah (US$ 5.6 million) and 67 trillion rupiah (US$ 7.5 million) had not been spent in Siak and Pelalawan districts respectively (Siak Forestry Agency Interviewee, 2009; Pelalawan Forestry Agency Interviewee, 2010). The officials of the forestry agencies in Siak and Pelalawan districts mentioned that the low spending of the reforestation fund was due to the fact that the guidelines provided by the national government cannot be fully applied at the local level. According to Ministerial Decree 14/2008, the uses of the reforestation fund transferred to the local level are as follows:

1. 60 per cent from the fund should be used for rehabilitation within the forest zone, while 40 per cent should be used outside the forest zone;

2. rehabilitation within the forest zone should be conducted within protection forests, managed by district forest agencies, and in production forests currently without active concessions;
3. rehabilitation outside the forest zone should be focused on water catchment, dams, lakes and water flows and river banks.

These criteria currently constrain the use of the fund, which leads to under-spending. An official noted that:

> The guidelines state that reforestation should be implemented inside and outside forest zones. Reforestation inside forest zones cannot be conducted within areas under active concessions. In reality, almost all forest zones in Pelalawan are currently within logging concessions. As a result, most of the activities can only be conducted outside forest zones, which leads to low spending (Pelalawan Forestry Agency Interviewee, 2010).

In 2010, a new ministerial decree was issued to revise the regulation, resulting in the introduction of two new features. First, when district governments have difficulties in finding a location within the forest zone, the use of the reforestation fund can be used according to local conditions. Second, the use of the reforestation fund outside the forest zone can also be allocated for rehabilitating peat land and peat swamp. However, as the interviews were conducted before the issue of the decree, the impact of this new regulation cannot be assessed.

The decree on the use of the reforestation fund also stipulates that activities financed by the fund should be conducted by a third party, such as local communities. However, in reality, local communities have a low capacity to implement rehabilitation activities. 'The reforestation activities need to be conducted by local communities, however, they are not technically ready to conduct forest rehabilitation' (Siak Forestry Agency Interviewee, 2009). Hence, building the capacity of local people is necessary before they can become actively involved in the rehabilitation activities. Furthermore, the national government stipulates standard prices, for instance, for seedlings and wages for every rehabilitation activity pursued at the local level. In reality, the standard prices do not reflect the local situations. Local governments faced problems in using the fund, as the standard prices were often too low and local communities refused to participate in the rehabilitation activities.

In Papua, too, officials from the Forestry Agency reported that the reforestation fund had not been used for rehabilitation activities, although for reasons that are different from those reported in Riau. As previously mentioned, the distribution of the reforestation levy is now labelled as a 'shared revenue'. This label misled local governments in Papua to consider the reforestation fund similar to other shared-revenues in the forestry sector, such as the forest licence fee and the forest resource rent. Shared-revenues are generally unconditional in nature and can be used for any activity or local priority approved by local parliaments during the budgeting process.

In both Jayapura and Sarmi districts, local governments perceive that use of the reforestation fund is unconditional. Hence, the use of the fund is decided by the local parliaments each year. If the forestry agency intends to use the fund for rehabilitation, the head of the forestry agency should propose and negotiate this during the budgeting process with the district head and local parliament. A local forest official in Jayapura district suspected that the fund cannot be used because the head of the forestry agency is often reluctant to negotiate the use of the fund for forest rehabilitation with the head of the district, as it could potentially create tensions with his superior. Forest officials in Sarmi district revealed a similar situation. When the finance agency in Jayapura district was asked about this issue, the following explanation was provided:

> The revenue sharing from the forestry sector is transferred in one package including the forest resource rent, the forest license fee and the reforestation fund. We do not know the amount of each fund as only forest officials know about it. As soon as the transfer is received in the local account, we treat them all as one package. We don't consider the source of transfers anymore and we treat them all as general funds. The spending of general funds should be negotiated in the local budgeting process. So, in the distribution of funds, we no longer consider the source of funds (Jayapura Finance Agency Interviewee, 2010).

As a result, the reforestation fund transferred to the local level cannot be automatically used for forest rehabilitation in Papua, although the guidelines provided by the national government state that the fund should be spent on forest rehabilitation, with clear criteria from the national government. Local governments have different interpretations of the guidelines provided by the national government, which impede the use of the fund.

The existing situation of IFT implementation, particularly in the forestry sector, emphasises the importance of considering local characteristics and situations in designing IFTs to finance conservation. Since local capacity in public financial management varies between regions, the guidelines related to the funds transferred by the national government need to take into account the voice of local governments as the recipients and the implementers.

5.4 CONCLUSION

To summarise the views on the design of IFTs, government officials' preferences on conditionality of IFTs for conservation vary depending on the mandates of the institutions they work for. For instance, finance agencies prefer a conditional transfer with clear guidance from the national government on how the funds should be spent, whilst technical agencies favour a revenue-sharing mechanism that carries more flexibility in terms of spending at the local level. Related to the distribution formula, however, there is a consistent pattern of responses provided by local governments, which argue for a higher portion of the revenues to be allocated to the local level. The responses provided by interviewees on their preferred mechanisms also consider the existing capacity of local governments in managing public resources transferred to the local level.

The implementation of conservation would have impacts on local governments' fiscal capacity – both on the expenditure and the revenue streams. On the expenditure side, local governments require sufficient financial resources to perform the devolved roles and responsibilities in conservation, previously discussed. Moreover, as forest exploitation and land-use change generate revenues, conservation could limit the revenues of local governments because of the restrictions on pursuing productive activities in forestlands (see Chapter 7). Impacts on the expenditure and revenue streams of local governments would then reduce local governments' capacity to deliver public services.

Considering the impacts of conservation on the expenditure and revenue streams, IFTs to finance conservation need to serve two important purposes. First, IFTs should ensure that conservation does not affect the fiscal capacity of local governments to deliver public services to their citizens. Second, IFTs need to finance specific services to be performed by local governments to support the successful implementation of conservation. The first objective of

IFTs for conservation is to ensure that sufficient compensation is provided for local governments to offset the opportunity costs from the forgone benefits due to restrictions on pursuing alternative land-use activities. The second objective is to transfer resources to finance the management of conservation areas at the local level, hence, it deals mainly with the management costs of conservation.

The existing shared-revenue (SR) mechanism in Indonesia could be appropriate for conservation if it is considered as a payment for environmental services provided by forest conservation. The payment for environmental services, such as from REDD+, can be considered similar to revenues generated from other forest commodities. As discussed earlier, an SR mechanism is usually utilised to distribute taxes and revenues from natural resources extraction collected by the national government to the producing regions. Although an SR could be considered as an unconditional transfer because the funds can be spent at the local level without restrictions, the distribution formula for an SR is based on the total taxes and fees collected from goods or services produced by a particular sector at the local level. The existing formula for the SR in Indonesia can accommodate the insertion of services generated by forest conservation. In the case of REDD+, for instance, the carbon emission reductions can be used as considerations in developing the formula. In the implementation of REDD+, a local government could be considered to be eligible on the basis of its participation in a REDD+ scheme and on the achievement of emission reductions from the forestry sector in the locality.

The Indonesian conditional grant (CG) mechanism has the potential to be used to distribute REDD+ revenues. As previously mentioned, CGs finance national priorities and commitment at the local level. If local governments need to perform activities or forest-related services to support the implementation of REDD+, the national government needs to transfer the necessary financial resources. As previously discussed, without providing sufficient resources, local governments would have no incentive to provide forest-related services due to the issue of spatial externalities. The goals of CGs include the achievement of minimum service standards and the correction of spatial spillovers (Usman et al., 2008). The existing design of the Indonesian CG also allows the insertion of REDD+ considerations into the technical criteria. For instance, the technical criterion for the existing CG for forest rehabilitation is the degraded forests/lands and degraded mangroves. In the case of the distribution of REDD+ revenues, the technical criteria for the CG could be the costs of providing REDD+ related services at the local level. Finally, the eligible regions are determined taking into consideration both general and specific characteristics, including fiscal capacity and regional characteristics.

The Indonesian unconditional grant (UG) mechanism is inappropriate for conservation as it aims to finance routine spending at the local level and to address the imbalance between expenditures and revenues generated locally. As noted earlier, the source of the UG is the national government's net income and a major portion of the transfer finances the civil servants' wage bill at the local level. If the distribution of finance for conservation were to be transferred to local governments using the UG, it would require integrating environmental services provided by conservation into the distribution formula of the UG. Integrating such services into the transfer could therefore distort the original purpose of the UG transfer and the transfer cannot be directly linked to policies or activities of conservation at the local level. Another drawback of using the UG to transfer REDD+ revenues is that the source of the UG is not directly linked to revenues generated by a specific sector, nor is the allocation of UG intended to finance a specific public service at the local level. A change in the regulation would be required, since the existing distribution formula does not allow the insertion of technical indicators linked to a particular sector or public service.

NOTES

1. ICMS – Imposto sobre Circulação de Mercadorias e Serviços.
2. GERHAN is *GerakanNasionalRehabilitasiHutandanLahan* or the national movement for land and forest rehabilitation. The Ministry of Forestry implements GERHAN through local bureaus of watershed control and forest resources conservation in collaboration with the provincial and district forestry services (Barr et al., 2009).

6. The design of REDD+ and decentralised forest management

As discussed in the previous chapters, local governments often have a strong interest in exploiting forests to generate revenues from taxes and fees as well as from the distribution of relevant national taxes and fees. REDD+ measures may restrict local governments from pursuing productive activities in forests and reduce this revenue stream which is needed to finance local public services. Local governments would also need to perform a number of activities to address locally specific causes of deforestation to ensure successful implementation of REDD+.

Meaningful participation of local governments in REDD+ implementation is imperative. Successful implementation of REDD+ requires not only the transfer of funds, but it also requires addressing governance challenges to its implementation (Corbera et al., 2010; Ebeling and Yasué, 2008). Most countries responsible for emissions from deforestation and forest degradation have implemented certain forms of decentralisation in public administration and forest management. Appropriate distribution of revenues among government levels needs to be carefully considered to compensate the costs of REDD+ implementation at the local level.

In this chapter we discuss options for fiscal instruments to distribute REDD+ revenues by drawing on the aspects, experience and issues concerning IFTs that have been addressed in previous chapters. We begin by considering key aspects of the progress of REDD+ negotiations in the UNFCCC and possible REDD+ implementation models. The discussion is framed within the context of decentralised forest management. We then provide the perspectives of local officials about the possible implementation models for REDD+ in Indonesia.

6.1 REDD+: THE CONCEPT'S EVOLUTION AND DESIGN ELEMENTS

The concept of REDD+ has evolved significantly since it was first discussed as the 'Compensated Reductions' scheme. Santilli et al. (2005) suggested the need to provide compensation for reducing deforestation using the historical deforestation rate as the baseline scenario. The concept of REDD was quickly accepted as a cost-effective measure, particularly after the Stern Report (2006) estimated that the cost of forest protection in eight countries responsible for 70 per cent of emissions from deforestation and land degradation will be around US$ 5 billion per annum initially, with increasing marginal costs over time.

Since the initial discussion of REDD+, a number of issues have been foreseen as a major challenge to the successful implementation of reducing emissions from land-based sectors. Two main technical challenges of REDD+ are leakage and non-permanence. Leakage occurs when forest protection in one location can cause deforestation, or adds more pressure to forests, outside the boundary of a certain project or country (Schlamadinger et al., 2005; Meridian Institute, 2009; Myer, 2007). Carbon sequestered in forests is often considered to be non-permanent because carbon can be released into the atmosphere in the future due to human or natural disturbances (Schlamadinger et al., 2005; Meridian Institute, 2009; Myer, 2007).

There has been much development since the concept was first acknowledged in the Conference of the Parties (COP) 13 of the UN Framework Convention on Climate Change (UNFCCC) in Bali, Indonesia, in 2007. A number of proposals were put forward about the design of REDD+ in the international negotiations that addressed the major challenges identified. The concept has also expanded since 2007. In the Copenhagen COP 15 in 2009, the parties added the need to include three more activities to be eligible under the mechanism, including: (i) conservation of carbon stocks; (ii) sustainable management of forests; and (iii) enhancement of carbon stocks. The agreement reached at the 2010 Cancun meeting of the UNFCCC on REDD+ was expected to result in a significant flow of funds from developed to developing countries. Finally, the 2013 COP provided detailed guidance on important elements that should be established by participating countries to implement REDD+.

The COP 19 held in Warsaw in 2013 agreed on the work programme to progress towards the full implementation of REDD+. Several other decisions made in Warsaw include: the need for national institutional arrangements, modalities for national forest monitoring systems, safeguards, forest reference emission levels and/or forest reference levels and modalities for measuring, reporting and verifying (UNFCCC, 2013). To understand how the key elements of the design would influence the implementation of REDD+ in participating countries, this section discusses the elements that are closely related to the intergovernmental fiscal transfers.

6.1.1 Reference Emission Levels

The approach to setting reference emission levels has become one of the most debated issues surrounding the design of a REDD+ mechanism. The reference emission level is the level against which the impacts of REDD+ policies and measures are assessed to determine whether participating countries have reduced emissions and should receive financial rewards (Parker et al., 2008; Meridian Institute, 2009). A reference level or baseline provides an estimate of the future GHG emissions from deforestation and forest degradation that would take place in the absence of REDD. Once the reference levels are set, the crediting reference level will establish the level of emissions that are planned to be achieved.

One of the most challenging tasks in setting reference emission levels is to accommodate the different circumstances found within developing countries. A REDD+ mechanism needs to be attractive enough for countries with high rates of deforestation as well as for those with low rates to participate. The greater the number of countries participating in the REDD+ mechanism, the greater the expected reduction in international leakage (Santilli et al., 2005).[1] Mollicone et al. (2007) recommend that the global average deforestation rate be used as a benchmark to accommodate countries with a range of deforestation from low to high rates. Under this approach, host countries with deforestation rates above the global average would be compensated for a reduction in their national deforestation rates during the commitment period as compared to the pre-commitment period. Countries with past deforestation rates lower than the global average would be rewarded for not increasing their deforestation rates higher than the pre-commitment level. This proposal also adds the element of forest degradation into the calculation of reference emission levels.

Progress has been made in the UNFCCC negotiations regarding the development of reference emission levels (REL). The Conferences of Parties 16, 17 and 19 decided that REL should be applied at the national level. National reference emission levels need to be developed based on national factors such as: historic emissions and removal rates, forest cover, expected future trends and the capacity for emission reductions such as GNP per capita (UNFCCC, 2009a). Participating parties can also develop sub-national REL that are consistent with national level REL. REL should be reported biannually to the UNFCCC. The COP 19 decided that countries should ensure the following when developing their REL (UNFCCC, 2013):

1. maintaining consistency with corresponding anthropogenic forest-related greenhouse gas emissions by sources, and removals by sinks, as contained in the national greenhouse gas inventories;
2. taking into account historical forest data;
3. ensuring transparent, complete, consistent and accurate data by providing methodological information, description of data sets, approaches, methods, models and assumptions used;
4. providing a description of relevant policies and plans to reduce deforestation and forest degradation;
5. describing changes to previously submitted REL;
6. providing the definition of forest used in the construction of the REL and, if it is different from the one used in the national greenhouse gas inventory or from the one reported to other international organizations, why and how the different definition was chosen;
7. including assumptions about future changes to domestic policies in the construction of REL.

Due to wide variations in regional situations across a country, local reference emission levels would vary from one locality to another depending on, inter alia, the total forest area, opportunity costs and the capacity to implement policies and measures at the local level. In order to develop national and sub-national reference emission levels, an analysis of land-use change patterns at the local level is necessary. Macroeconomic models, which are often considered the appropriate approach to forecast national reference emission levels, fail to take into account the causes of deforestation originating from land-use changes triggered by local factors (Bird, 2005). Even when the drivers of deforestation are identified, predictions that use national models are of limited use in understanding the strength of the drivers, the influence of the drivers across time and space, and the interrelationship between the drivers. Understanding the social processes influencing the decision to pursue land-use change at the local level is necessary to assess forest cover changes at the national level (Mascia et al., 2003; Dalle et al., 2006).

In developing REL, countries should also consider multiple drivers of deforestation and forest degradation. During the COP 19, the UNFCCC parties emphasised the need to acknowledge that livelihoods may be dependent on activities related to drivers of deforestation and forest degradation; hence, addressing these drivers may have an economic cost and implications for domestic resources. Actions to address these drivers should therefore be unique to countries' national circumstances, capacities and capabilities. The involvement of the private sector has been highlighted as an important element to reduce the drivers of deforestation and forest degradation (UNFCCC, 2013).

6.1.2 Monitoring, Reporting and Verification

Forest monitoring, reporting and verification (MRV) is an important activity for REDD+ because it will determine whether a country has achieved significant and credible reductions in emissions and enhancement of carbon stocks. Following such processes, the financial benefits can be granted to participating countries based on performance. The monitoring and reporting system requires national coordination and a cooperative mechanism to link forest carbon MRV and national policy for REDD+ (Herold and Skutsch, 2009). A national body would collect and analyse data based on the stipulated protocols to estimate emission reductions or enhancement of carbon stocks at national and sub-national levels. Two variables to be measured and estimated under the REDD+ scheme are: (i) forest area change (deforestation and re-growth of forest); and (ii) carbon stock change estimation (emission factors) (UNFCCC, 2009b). Data and information gathered should be well documented and reported to a designated international body based on the Intergovernmental

Panel on Climate Change (IPCC) methodologies and in compliance with UNFCCC guidelines. Finally, an independent framework is required to verify the long-term effectiveness of REDD+ actions at different levels and by different actors (UNFCCC, 2009b).

REDD+ funding would be granted based on performance. Measurable, reportable and verifiable indicators need to be developed for the implementation of REDD+ policies and measures. Subject to *ex post* verification, upfront financing may also be granted based on spending plans and stated commitments (UNFCCC, 2009a). After receiving compensation, distribution of REDD+ benefits among all stakeholders, including indigenous peoples and local communities, should be fair, efficient, transparent and equitable (UNFCCC, 2009a).

In order for the participating countries to obtain and receive result-based payments, a report on the progress of REDD+ should be submitted biannually to the UNFCCC. The report consists of transparent data and information related to the sources of forest emissions and absorption, forest stock carbon and the change. The report should be consistent with the REL and must (UNFCCC, 2013):

1. use tonnes of carbon dioxide equivalent per year (CO2 eq);
2. report REDD+ activity or activities included in the forest reference emission level and/or forest reference level;
3. include information regarding territorial forest area covered;
4. report the date of REL and the date of the report;
5. include the period (in years) of the assessed REL.

6.1.3 Financing Options

Participating countries in a REDD+ scheme would be eligible for financial benefits when they reduce emissions. The financial benefits would need to be further distributed among all stakeholders, including indigenous peoples and local communities, efficiently, transparently and equitably (UNFCCC, 2009a). In order to determine how the revenues should be distributed to different stakeholders within a country, it is important to understand the distribution of costs across groups (Pagiola and Bosquet, 2008). As different land users face different costs in reducing emissions, the distribution of incentives could then be based on the marginal costs of reducing emissions (Boucher, 2008; Wertz-Kanounnikoff, 2008; Cattaneo, 2008; Strassburg et al., 2009). Any given producer incurs costs that vary according to the amount produced (Boucher, 2008). Some producers could reduce emissions inexpensively, while for others it would be more costly.

Several proposals have been put forward about the design of a REDD+ mechanism at the international level. For instance, Strassburg et al. (2009) and Cattaneo (2008) propose the creation of a global pool of all revenues generated for REDD+. Participating countries would receive a share from the global pool only when there is a positive difference between global reference emission levels and total emission reductions achieved by all participating countries, as a group, within a given time frame. The distribution of incentives, at a minimum, should compensate the costs of REDD+ implementation, including opportunity costs and management costs, as well as transaction costs (Cattaneo, 2008; Boucher, 2008; Strassburg et al., 2009). As the price of carbon in the market is expected to be higher than the costs of REDD+ implementation, the net benefits of REDD+ would be the difference between the payments for emission reductions and the costs of achieving the reductions (Boucher, 2008; Cattaneo, 2008; Meridian Institute, 2009; Pagiola and Bosquet, 2008). The distribution of incentives to participating countries could be based, for example, on the marginal costs of reducing emissions, the standing carbon stocks sustained in the countries, or a combination of these two methods (Cattaneo, 2008; Strassburg et al., 2009). In decentralised countries, local governments could then be requested to make bids for unit reductions or, alternatively, all local governments could be paid the same amount per tonne of reduction, irrespective of the costs.

The flow of funds could depend on the financing options approved in the international REDD+ negotiation. The financing options to implement REDD+ can be categorised as either market or non-market approaches. Market approaches enable developing countries to generate credits from REDD+ measures and sell them to Annex 1 (developed) countries of the Kyoto Protocol, which may purchase and use the credits to meet their emission reduction commitments. Non-market or fund-based approaches propose a fund created by Annex I countries to reward developing countries for their efforts to reduce emissions from deforestation and forest degradation (Johns et al., 2008). The COP of UNFCCC has also decided that the

Green Climate Fund will include REDD+ financing. The Green Climate Fund is an operating entity of the Financial Mechanism of the UNFCCC that supports projects, programs, policies and other activities in developing country Parties.

Several studies have discussed the options for REDD+ financial mechanisms at the national level (Eliasch Review, 2008; Vatn and Angelsen, 2009). The Eliasch Review (2008) suggests two possible mechanisms to distribute REDD+ funds from the national to sub-national levels: (i) national coffers, where the revenue is treated as any other sovereign revenue; or (ii) a special fund, which is earmarked to finance forestry or non-forestry programmes. The second option allows the involvement of international actors, as the funds would be managed outside the state administrative system. Vatn and Angelsen (2009) further discuss four possible mechanisms to channel the flow of funds from the national to sub-national levels (see Figure 6.1):

1. project-based fund, which is carried out under project-based implementation; hence, revenue distribution is directly channelled to the project areas;
2. separate national fund, which is established outside state administration and governed by a board of representatives and stakeholders;

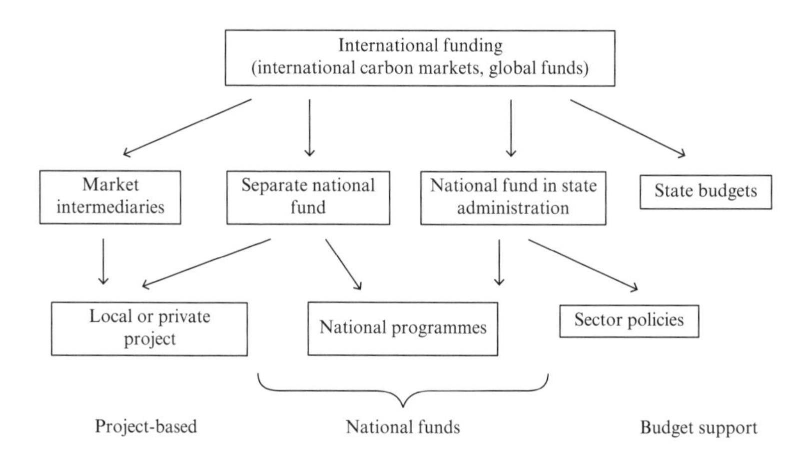

Source: Vatn and Angelsen (2009, p. 64)

Figure 6.1 Options for national REDD+ funding architecture

3. national fund within state administration, which is administered by an independent board with members from the relevant states and public administrations and possibly civil society – the board can allocate money to specific programmes, sector administrations and individual projects;
4. specific budget support, which is channelled through existing budget systems in the form of general budget support or earmarked funding.

These studies assume that, irrespective of the agreed global financing, national governments could choose one of these mechanisms (Figure 6.1).

6.2 SUB-NATIONAL IMPLEMENTATION AND JURISDICTIONAL REDD+

The concept of a jurisdictional REDD+ has been advocated where reference levels should be developed at the jurisdictional level. A jurisdiction is a government administrative area where a particular system of laws is applied. A jurisdiction is usually led by an authority that has the power or right to govern and to interpret and apply the law. A jurisdiction usually has a common goal, which is the prevailing wealth in the geographical area (Feldman and Martin, 2005).

The central feature of jurisdiction-wide REDD+ programmes is the reference level that is developed for an entire jurisdiction. Deforestation and/or GHG emission targets can be established and legally adopted together with a definition of the reference level. The target provides the state with a specific goal that can facilitate planning programmatic investments, and future evaluation of programme efficacy (Nepstad et al., 2012). Jurisdictional REDD+ includes the concept of a nesting approach, which is a system that reconciles REDD+ programmes of a jurisdiction (e.g. provincial) with the higher level (e.g. national), and reconciles with REDD+ programmes from the lower level (e.g. district). The ultimate goal is to ensure that emission reductions 'add up' at the jurisdictional level, whether national or sub-national, with each participant receiving proper credit for its contribution (Nepstad et al., 2012).

In order to ensure the successful implementation of REDD+ in decentralised countries, it is important to consider which tasks could be devolved to what level in these countries. The basic principle of subsidiarity in decentralised public administration is that 'tasks and powers should rest at the lower level subunit unless allocating them to a higher central unit would ensure higher comparative efficiency or effectiveness in achieving them' (Føllesdal, 1998, p. 190). In the implementation of REDD+, several activities are best handled by the national government, while others would be best devolved to the local level. Based on the ongoing negotiations, national governments may need to develop national carbon accounting, monitor the implementation of REDD+ policies and measures, receive and distribute REDD+ credits and assume liability after the payment has been received. This study suggests that local governments are in a better position to develop local policies and measures at the local level. Local authorities have better specific information related to local resources, which results in better-targeted policies and lower transaction costs (Ribot et al., 2006). Several benefits of having local governments involved in the implementation of REDD+ can thus be summarised:

- to tackle the specific causes of deforestation at the local level, as the drivers vary from one location to another within a country depending on the economy and the needs of local people;
- to ensure greater participation of sub-national groups in the decision-making process where the decision-making process of land-use has been devolved;
- to increase the efficiency of REDD+ implementation through internalising costs and reducing transaction costs.

Two main activities that should be carried out by local governments are the development of reference levels and the facilitation of the development and implementation of REDD+ programs in their jurisdiction. The latter activity should be seen as an integral part of the development of the reference levels, as it would inform decisions about much to reduce in terms of emissions and how that is going to be achieved.

6.3 FROM THEORY TO PRACTICE: THE DESIGN OF REDD+ AT THE NATIONAL DOWN TO THE LOCAL LEVEL

Having discussed the general features of the concept and design of REDD+, we now turn to Indonesian government officials' perspectives about the possible models for the implementation of REDD+, particularly the distribution of benefits. The discussion in this section is based on the interview data with officials that were carried out in 2009, before the REDD+ dialogue started at the national level. The officials whose views are reported here were the same officials interviewed in relation to the previous chapter. The implementation status of REDD+ to date will then be discussed in the next section.

Five national and local officials thought that if REDD+ was going to be implemented through a global fund, the national government should be entirely responsible for the implementation of the scheme. Under this model, reducing deforestation and forest degradation is perceived as part of the national commitment to an international agreement. In return, the national government would receive financial benefits to support its efforts to implement REDD+. The financial flow could then be managed under a grant or a loan mechanism.

> Revenues from REDD+ compensate our commitment to forest conservation. The developed countries would provide the support in the form of grants or loans. If this were the case, then the mechanism of international grants and/or loans regulated by the law would need to be implemented (Finance Ministry Interviewee #1, 2009).

Five other national and local respondents believed that a market should be created to trade carbon credits generated from REDD+. Under this mechanism, the national implementation of REDD+ could be performed either by a public or private entity. One official from the Ministry of Finance believed that if forests were under active concessions, then it would be the responsibility of the private sector to implement REDD+. However, it would be the responsibility of the government to implement REDD+ in state forests, without active concessions. In contrast, an official from the Ministry of National Development Planning thought that the state should not be involved in business or profit-making activities because it does not have a comparative advantage:

> The government is not supposed to be involved in carbon trading. The government can obtain benefits from taxes. But to ensure efficiency, the national government should not be involved in trading activities. If they want to be involved, then they can form state-owned companies at the national and local levels (National Development Planning Ministry Interviewee, 2009).

At the district level, all respondents from the forestry agencies suggested that the implementation of REDD+ could be 'outsourced' to a third party, including private, non-government organisations or local communities:

> Local governments do not need implement REDD+. Local governments should only conduct facilitation and programming. The implementation can be outsourced to a third party (Siak Forestry Agency Interviewee, 2009).

The rationale behind the preferences of the respondents from the forestry agencies is: (i) local governments should play the role of a regulator that issues the necessary guidelines and monitors implementation (Siak Forestry Agency Interviewee, 2009); (ii) local governments have no (or low) capacity to implement REDD+ (Siak Forestry Agency Interviewee, 2009; Pelalawan Forestry Agency Interviewee, 2010); and (iii) local communities can obtain direct benefits from the implementation of REDD+ (Sarmi Forestry Agency Interviewee, 2009; Jayapura Forestry Agency Interviewee, 2010).

Regardless of the mechanism agreed at the international–national level, all government officials agreed that local governments should hold a number of responsibilities to ensure the successful implementation of REDD+. These responsibilities are mainly related to ensuring that local people gain the most benefits from REDD+ activities, and monitoring and evaluation of REDD+ implementation. Consistently, all respondents emphasised that local governments should ensure that the needs of local people are taken into account at all implementation stages of REDD+. Specific roles and activities that local governments should play in the implementation of REDD+ include:

- ensuring local community welfare – i.e. alternative livelihoods, village infrastructure and community development (Riau Provincial Forestry Agency Interviewee #3, 2009);

- developing community forestry programmes, strengthening governance at the community level and rehabilitating critical lands (Papua Provincial Forestry Agency Interviewee, 2010);
- monitoring and evaluation of REDD+ implementation (Pelalawan Forestry Agency Interviewee, 2010);
- facilitating, programming, monitoring, providing guidelines and building community forestry, assisted by a third party (Siak Forestry Agency Interviewee, 2009).

Government officials' perspectives on the national design of REDD+ implementation vary depending on the mechanism to be agreed in the international negotiations. If the financial benefits of REDD+ were to be transferred through a global fund, REDD+ was perceived to be a scheme created to support the efforts of developing countries to reduce emissions under a climate change convention. Respondents thought that this implementation mode placed more responsibility on the national government to carry out REDD+ at the national level. Financial compensation would be considered similar to foreign grants that would be distributed from the global to a national fund. In contrast, if REDD+ were implemented through a market mechanism, the scheme was considered to be an economic activity, similar to alternative land-use activities to REDD+. This view, consequently, shifts the responsibility from the public sector to private entities because, in the belief of the government officials interviewed, the state should be less involved in market activities. The state can obtain revenues from REDD+ from taxes or fees, as is already the case with other economic activities.

In the case of the global fund model, REDD+ national implementation would need to use the foreign grant modality. According to Government Regulation 02/2006, the foreign grant modality is to be implemented when the country receives funds from bilateral or multilateral donors. Under the foreign grant modality, a project management unit (PMU) administers the funds outside the state treasury, although use of the funds will need to be reported in the state budget document. The use of a foreign grant is strictly earmarked for specific activities predetermined in the grant agreement document. Once the foreign grant agreement is signed, sectoral ministries are then responsible for administering the project.[2] Foreign grants can also be channelled to a Public Service Unit (PSU), where the relevant ministers or district heads/governors take full responsibility for a PSU's management.

If REDD+ were implemented under a market mechanism, stakeholders perceived it to be a profit-making activity, similar to natural resource extraction or international trading. In the case of natural resource extraction, for instance, the state can invite private companies or establish state-owned enterprises to manage the resource and trade the products (Searle, 2007). Under a market mechanism, the state is not usually involved in profit-making activities and dealing directly with international buyers. Rather, it should create a conducive environment by issuing necessary regulations to encourage the participation of the private sector (Box, 1999; Parsons, 1995; Bradbury and Waechter, 2009). An official from the Ministry of National Development Planning argued that governments lack the comparative advantage to be involved in profit-making activities, and that if the state decides to become involved in REDD+ implementation, a state-owned enterprise should be established to deal with international buyers and carry out REDD+ implementation. Moreover, most respondents at the local level considered the need to outsource the implementation of REDD+ to third parties such as local communities, private entities or non-government organisations.

Understanding government officials' perspectives about the roles and responsibilities of different actors in the implementation of REDD+ is important before determining the appropriate mechanism to distribute REDD+ revenues. Government officials interviewed at all levels generally believed that more authority was required for local governments in forest management, particularly in REDD+ implementation. Although providing local governments with greater power does not always lead to better forest management and conservation (Tacconi, 2007), local stakeholders need to be allowed to pursue policies according to local people's needs (Cheema and Rodinelli, 1983). Since REDD+ would require additional forest conservation that may not necessarily yield additional environmental services for local residents, local governments need to be provided with the authority to decide their participation in a REDD+ scheme on behalf of their constituents. Assessments of the costs and benefits of REDD+ and local target measures of emission reductions need to be conducted, taking local situations into account. In the implementation of REDD+, local governments can be expected to work better together with the national government through voluntary interactions because of mutual benefits and similar goals and objectives (Oliver, 1990; Levine and White, 1961; Parsons, 1995).

After a locality decides on its participation in the implementation of REDD+, local governments can decide to outsource implementation activities to a third party. Respondents interviewed at the district level, particularly from the forestry agencies, suggested that REDD+ could be outsourced to private companies, non-government organisations or local communities. Private companies could be invited to develop REDD+ projects and also perform necessary transaction (including dealing with buyers) and management activities. Respondents also argued that the implementation of REDD+ can be outsourced to forest community groups to allow local people to obtain direct benefits from REDD+. Although many REDD+ activities could be outsourced, respondents also believed that local governments would still be required to provide a number of forest-related services to ensure successful implementation of REDD+, including issuing the necessary guidelines, monitoring and evaluation, and forest protection. Furthermore, the state, including local government, has an obligation to ensure that local people obtain the benefit they deserve from the economic activity (Box, 1999; Bradbury and Waechter, 2009).

The difference between the national designs of REDD+ implementation discussed above is how government stakeholders would obtain REDD+ benefits. The foreign grant modality might not provide the flexibility to use the fund for general budget support for government. Foreign grants are usually earmarked to finance specific activities, which can therefore be allocated to finance REDD+ management activities within a specific sector or at local levels. For instance, the government could implement an agricultural policy financed by foreign grants, however, the flow of funds would not be channelled to the agricultural ministry budget, but would be disbursed through a PMU or a PSU. In contrast, under a market mechanism, the government can collect taxes or fees from the transaction of REDD+ credits in the market. Taxes and fees can then be used to provide general budget support and finance policies through the state budget. The revenues collected through taxes and fees can also be transferred to local governments using IFTs.

6.4 THE STATUS OF REDD+ IMPLEMENTATION IN INDONESIA

To date, most activities In Indonesia have focused on building institutions and dialogues for readiness to move towards a full implementation of REDD+, with minimum progress made on real actions to reduce deforestation. Numerous projects at the national or local levels have been supported by bilateral donors, non-government organizations and government organizations. In 2010, Indonesia signed a $1 billion deal on REDD+ with the Norwegian Government. Thirty million dollars were disbursed shortly after the signing and have not been entirely spent. Most spending has so far been on building REDD+ readiness for implementation.

Several REDD+ voluntary projects have been initiated since 2009. In Indonesia, these project-based activities are mostly operating under a restoration ecosystem permit issued by the Ministry of Forestry. The permit is usually issued to restore a logged-over area, following commercial logging activities, back to its original state. The Ministry of Forestry, through Decree 5040/2013, has already allocated an area totalling 2.7 million hectares for restoration ecosystem permits. As of December 2013, around 47 proponents have submitted their proposals to obtain a permit, while only 12 permits have been granted for a total area of 480,093 ha (Ministry of Forestry, 2014). There has been no specific decision regarding how carbon credits that may be generated by these projects will be included in the national or subnational reference emission levels.

Indonesia established a body dealing with REDD+ through the issuance of Presidential Decree 62/2013. However, the REDD+ Agency was merged with the Ministry of Forestry and Environment in 2015. The new structure is currently being finalised. The Agency had the following roles:

- develop a REDD+ strategy;
- develop social, environmental and financial safeguards;
- coordinate the development of REDD+ policies and their implementation;
- prepare and coordinate the REDD+ financial mechanism, as well as manage REDD+ funds;

- develop standards and methodologies for emission measurement from project and programme activities to consolidate and report as part of GHG emissions;
- increase the capacity of ministries and agencies;
- provide recommendations at the international level;
- coordinate law enforcement for REDD+;
- facilitate conflict resolution;
- carry out the monitoring and evaluation of REDD+.

Presidential Decree 62/2013 on REDD+ Agency did not specify the role of the Agency in developing REL, while it specifically mentioned its role to conduct monitoring, reporting and verification activities. The Ministry of Forestry developed the REL for the forestry sector through the issuance of Ministerial Decree 633/2014. The levels of the REL included in the Ministerial Decree, however, differ significantly from those included in the Second National Communication (SNC) to the UNFCCC, which were used to determine the reference level of Presidential Decree 61/2011.[3]

The REDD+ Agency also proposed to set up a trust fund to receive and distribute REDD+ benefits in Indonesia – Fund for REDD+ in Indonesia, or FREDDI. The proposed design of FREDDI would manage REDD+ fund through a trust fund with several financing modalities:

- grants, both large scale and small scale;
- investments, i.e. funding in the form of equity and loans;
- performance-based payment – that is, financing in exchange for achieving results-verified REDD+.

Although the REDD+ Agency mentioned that the main beneficiaries of REDD+ benefits would be government actors, there has been no clear proposal about whether REDD+ finance would be channelled through the budgetary system or not. Channelling funds outside the budgetary system would mean that they are not included in the annual national budget law and may not be subjected to the same level of scrutiny and accounting standards as the government budget. Using the existing government system would be preferable because it is important to ensure that REDD+ activities implemented by the government were consistent with national development priorities. This would not require a revision of the existing regulations, which would prolong the timeframe for REDD+ implementation.

Distribution of REDD+ revenues from the national to the local level could utilise IFTs that are commonly applied within decentralised countries. IFTs could be used to distribute REDD+ revenues vertically to local governments or to compensate for the spatial externalities of conservation. REDD+ credits can be considered as a forest commodity that can be traded in the market, resembling other forest products such as timber. As with the revenues generated from timber extraction, REDD+ benefits could then be returned to the regions that generate carbon credits, using a revenue-sharing mechanism. Moreover, forest conservation can be viewed as a public service provided by local governments to support the national commitment. Forest conservation creates spatial externalities because it involves local opportunity costs, but it generates benefits far beyond municipal boundaries (Ring et al., 2010). If local governments were compensated for the spatial externalities of forest conservation, they could be expected to provide a sufficient area of forests, including setting aside a significant quantity of land under their jurisdiction for conservation.

6.5 CONCLUSION

National governments could devolve or delegate some of the implementation tasks of REDD+ to subnational actors. REDD+ implementation activities that could be devolved to local levels include: developing baselines, identifying REDD+ targets and measures, and monitoring REDD+ implementation and/or project development. Local governments could ensure greater participation and consider local specific situations and causes of deforestation in developing REDD+ baselines, setting targets for carbon reductions and identifying innovative REDD+ measures (Irawan and Tacconi, 2009; Larson and Ribot, 2009). As REDD+ measures would limit the revenue stream of local governments due to restrictions on pursuing productive activities in forestlands, meaningful participation of local governments in REDD+ implementation is important. Local governments should be allowed to assess the social costs and benefits of REDD+ implementation. They could then decline to participate if the REDD+ mechanism was not considered beneficial for their localities. Participation of local communities in the development of REDD+ strategies or policies is possible when the planning process is conducted at the lowest governmental level. Local stakeholders, who would be directly affected by REDD+ policies and measures, are often geographically distant from national authorities. When the planning process is devolved to the local level, local voices and socio-economic conditions are more likely to be taken into consideration in the development and implementation of REDD+. Furthermore, local governments have a comparative advantage to perform monitoring and law enforcement due to their close proximity to forests and to the local communities living in or near surrounding forests.

NOTES

1. International leakage occurs when the implementation of REDD policies and measures in one country causes an increase in emissions from deforestation and forest degradation in another country.
2. Government Regulation No 2/2006.
3. The then president committed to reduce emissions by as much as 26 per cent using domestic resources and 41 per cent with international support in 2020. Using the SNC as the main reference, the emissions level would increase from 1.72 Gton CO2e in 2000 to 2.95 Gton CO2e in 2020 (Ministry of Environment, 2009). Based on the SNC, the target of emission reduction by 2020 was 0.767 Gton CO2e using domestic resources, where the forestry sector contributed to almost 80 per cent of the reduction. The REL set by the Ministry of Forestry would only reduce the contribution of the forestry sector to around 30 per cent of the total reduction of emissions.

7. Incentive structures influencing subnational governments' decisions on land-use change[1]

The previous chapter established that national governments should involve local governments in planning and implementing REDD+ activities and that IFTs could be used to distribute REDD+ revenues to local level governments to compensate them for the eventual forgone benefits of deforestation and forest degradation. In this chapter we assess the financial incentives currently influencing different stakeholders, particularly local governments, in pursuing land-use change and forest exploitation in Indonesia. The opportunity cost is considered to be the most significant indicator of those financial incentives (Pagiola and Bosquet, 2008). The estimation of opportunity costs carried out in this chapter will then be used in Chapter 8 to explain how to calculate the size of the IFTs that may need to be distributed to local governments.

Given their importance, research on the economics of REDD+ has given particular attention to the estimation of the opportunity costs of REDD+ and the required flow of funds at the global, national and sectoral levels (for example, Boucher, 2008; Butler et al., 2009; Grieg-Gran, 2008; Kindermann et al., 2008; Pagiola and Bosquet, 2008; Wertz-Kanounnikoff, 2008). Depending on the methods used, global reviews of the opportunity costs of REDD+ estimate them to lie between US\$2.51 per tCO_2 (Boucher, 2008) and a range of US\$10 to 21 per tCO_2 (Kindermann et al., 2008). The studies of opportunity costs have focused on private stakeholders – namely, companies and smallholders. However, the legal framework determining the rights over forests needs to be considered in estimating the opportunity costs of REDD+ and the associated incentives at the national level. The legal framework regulates who owns forests, who bears the costs of the implementation of REDD+ activities and, therefore, who should receive appropriate incentives to change deforestation-related behaviour (Gregersen et al., 2010). When the state claims ownership of forests – a situation common in many of the top deforesting countries (Agrawal et al., 2008; Tacconi et al., 2010) – the opportunity cost it faces in reducing emissions from forests is equal to the revenue stream it forgoes for not issuing permits for income-generating activities in forests (Gregersen et al., 2010). The costs and incentives faced by governments in the implementation of REDD+ should also be considered, including local level governments, as most countries responsible for emissions from deforestation and forest degradation have implemented some degree of decentralisation in public administration and forest management (Irawan and Tacconi, 2009). Therefore, this chapter estimates the opportunity costs of REDD+ accruing to the national, provincial and district governments and companies in the provinces of Riau and Papua. The estimation of government's opportunity costs is based on taxes, fees and charges generated by the alternative land-use activities, taking into account the implications of adopting alternative discount rates, an issue that has not been clearly addressed by previous studies focused on REDD+.

The methods used are presented in the following section, which reviews the literature on the estimation of the opportunity costs of REDD+ and on the discount rate, and then clarifies the data used. The results of the estimation of opportunity costs and their distribution are then considered, before discussing the implications for REDD+ and the conclusion.

7.1 METHODS

7.1.1 Opportunity Cost Analysis

The cost of implementing REDD+ includes opportunity costs, management costs and transaction costs. Opportunity costs are the benefits of the best alternative land use that are forgone as a result of reducing deforestation and forest degradation (Grieg-Gran, 2008; Pagiola and Bosquet, 2008; Wertz-Kanounnikoff, 2008). Management costs arise from activities such as the prevention of illegal logging, research for agricultural intensification and land titling to provide traditional and indigenous communities with incentives to safeguard forests (Nepstad et al., 2009; Pagiola and

Bosquet, 2008; Wertz-Kanounnikoff, 2008). Transaction costs relate to the processes to identify and negotiate REDD+ activities as well as to perform monitoring, reporting, verifying and certification of carbon emission reductions (Cacho et al., 2005; Milne, 1999). This chapter focuses on the opportunity costs, because they are thought to account for the largest share of the costs (Pagiola and Bosquet, 2008). It should be noted, however, that the other costs could be more than insignificant. The transaction costs reported in the literature are in the range of US$0.01–16.40 per tCO_2 (Wertz-Kanounnikoff, 2008), while the management costs, in the Brazilian Amazon for instance, were estimated at US$1–3 per hectare per year (Nepstad et al., 2009). This chapter focuses on REDD+ opportunity costs incurred by local governments, as the opportunity cost is the most important category of costs that provides a fair assessment of the causes of deforestation (Pagiola and Bosquet, 2008). Moreover, it is also problematic to predict the transaction costs faced by local governments at this point in time because the role of local governments in the implementation of REDD+ is still uncertain.

To obtain relatively reliable estimations of the opportunity costs of REDD+, two contextual issues need to be considered: the legal framework determining the rights of forest owners and the methodology used for the calculation of opportunity costs. The legal framework, which regulates who owns forests and what activities are allowed in forests, determine the real costs required to change the deforestation behaviour of forests' owners. 'In trying to understand the real costs required to change deforestation behaviour, it's important to start with the question of what rights the forest owner, or user has' (Gregersen et al., 2004, p. 4). When the removal of forest cover is permitted by law, the opportunity cost approach is a good approximation of the real costs of reducing deforestation incurred by a country. On the other hand, if deforestation is forbidden by law, the opportunity cost from the next best alternative of land use is not an appropriate measure, and the cost of improving law enforcement is the relevant cost (Gregersen et al., 2010).

Government stakeholders in Indonesia claim ownership of the majority of forest lands in the country. When forests are state owned, the opportunity cost it faces in reducing deforestation and forest degradation is equal to the revenue stream forgone by the state for not issuing permits for income-generating activities in forests (Gregersen et al., 2010). Companies who could obtain licences to operate in forests would also bear a portion of the opportunity cost from the forgone private benefits of commercial logging or plantation activities, but they do not necessarily have a right to compensation unless they have established rights to the forests in question.

Methods used in the calculation of carbon sequestration costs – including REDD+ opportunity costs – vary widely, resulting in a wide range of estimations. In their review of carbon sequestration cost studies since the 1990s, Richards and Stokes (2004) found that the research had: (i) made diverse assumptions about carbon yields, (ii) focused on distinct geographical areas and scales, and (iii) applied different time horizons and discount rates. They also observed that methods used by carbon sequestration cost studies can be further categorised into: (i) bottom-up engineering studies, which determine the value of inputs to production and derive the estimate of costs using observed prices from agricultural land rental; (ii) sectoral models, which account for the impact of landowners' profit-maximization behaviour to agricultural and land markets in the calculation of costs, by using spatial equilibrium models; and (iii) econometric studies, which analyse how landowners have historically allocated land use between agriculture and forests in response to differences in market prices (Richards and Stokes, 2004).

Approaches that are commonly used in the literature to estimate the opportunity costs of REDD+ are summarised by Wertz-Kanounnikoff (2008). The approaches include local as well as global empirical models and global simulation models. Local-empirical models estimate REDD+ opportunity costs using data collected directly within a confined or particular location, mostly through surveys. Based on the amount of carbon that would be lost should deforestation take place in the studied area, benefits accrued from the next best alternative land uses per-area ($/ha) are then converted to per-ton carbon stock ($ per tCO2eq). Local empirical models can be further aggregated to obtain global per-area opportunity costs of REDD+, which usually ignore the variation of carbon density across space (Wertz-Kanounnikoff, 2008). Finally, global simulation models, also referred to as dynamic partial equilibrium models, estimate costs incurred to global REDD+ suppliers using dynamic models of the world economy. The models often include important sectors that affect land use, such as the forestry sector, the agriculture sector and the energy sector (Kindermann et al., 2008; Wertz-Kanounnikoff, 2008). Local and global empirical studies usually apply the bottom-up engineering approach to estimate the forgone benefits from REDD+ alternative land use activities, while the global simulation models are similar to the sectoral approach discussed by Richards and Stokes (2004). As this study assesses the opportunity costs faced by different national stakeholders, it uses the local empirical approach.

Different land users face different costs and benefits in reducing emissions, and the cost of pursuing REDD+ can be described by a supply curve (Boucher, 2008). We will consider, therefore, the different land uses separately in order to identify their relative costs and benefits (and their distribution to the stakeholders) to assess whether they could be prioritised for cost-effective emission reductions.

It is important to emphasise that even when the same method is applied (for example, the local empirical method), the estimates of REDD+ opportunity costs – even those in the same country – can vary as a result of different data sources and, importantly, the discount rate used. For instance, the opportunity cost of oil palm in Indonesia has been estimated at US$9.85–33.44 per tCO_2 on mineral soil and US$1.63–4.66 per tCO_2 on peat soil by Venter et al. (2009), and US$7.66–19.24 per tCO_2 (without reference to soil types) by Butler et al. (2009).[2] The differences in estimates appear to be due to the sources of data and the assumptions used as follows.

- Profit data for oil palm plantations: Butler et al. (2009) assessed the profitability of a hypothetical oil palm plantation by calculating year-by-year yields and applying alternative pricing scenarios, whereas Venter et al. (2009) used profit data from published materials of several companies operating in Indonesia. The profit data used by Venter et al. (2009) were the net profits (after taxes) of companies operating in Indonesia. Butler et al. (2009) did not consider government taxes in their estimation of companies' net profits.
- Prices of palm oil: Butler et al. (2009) used price data and forecasts through to 2020 from The World Bank. They assumed that, under a high-price scenario, the price was constant at US$749 per ton from 2009 to 2039. Under a low-price scenario, they used the price of US$643 per ton in 2010, which decreases to US$510 in 2020 and remains at that level until 2039. Venter et al. (2009) did not state the price of palm oil used, as they use the net profit data from companies' reports.

- Carbon stock data: Butler et al. (2009) used an average amount of avoided emissions of 682.92 tCO_2/ha, while Venter et al. (2009) spatially estimated the avoided emissions in mineral and peat soil forests at 389 tCO_2/ha and 2,249 tCO_2/ha respectively.
- Discount rates: Butler et al. (2009) used a 10 per cent discount rate, whilst Venter et al. (2009) applied an 8 per cent discount rate.

7.1.2 Discount Rate

According to Smith (2011), the choice of the discount rate depends on whether: (i) the policy question is marginal or non-marginal; (ii) social or private preferences should be considered; and (iii) the country is developed or developing (that is, the relative level of income). A policy aimed at implementing REDD+ over large areas for a long period of time is a non-marginal policy. Non-marginal policies are evaluated using a social discount rate, rather than a market rate which reflects private preferences (Smith, 2011). Social discount rates are lower than private ones because, inter alia, social preferences place greater value on the welfare of future generations compared with private preferences. In relation to the development stage of a country, the Garnaut Review (2011) suggests that discount rates applied in developing countries should be higher than those in developed countries. This is due to the fact that the present generation in developing countries can be expected to be significantly poorer than future ones, whereas the gap in welfare between generations is more limited in developed countries.

The social discount rates used by major assessments of climate change policies, such as the Stern Review and the Garnaut Review, were as low as 1 to 2 per cent (Smith, 2011). However, the most common rate used by studies of the opportunity costs of REDD+ has been 10 per cent (Grieg-Gran, 2008). To account for the gap between the wealth of different generations in a developing country such as Indonesia, we use a social discount rate of 5 per cent, which is significantly higher than that used in the Stern and Garnaut Reviews. A positive rate (determined from observation of market data) could also be considered to approximate social preferences, with the most appropriate one being the yield of long-term government bonds (Smith, 2011). For Indonesia, in 2011 this rate was about 7 per cent.[3]

International practice recommends discount rates varying from

1 to 15 per cent in the assessment of the value of projects (Harrison, 2010). To enable consideration of the different social and business issues that arise in assessing land-use options, we present opportunity costs calculated using discount rates of 5, 10 and 15 per cent with a 30-year time horizon. The rate of 10 per cent is slightly higher than the rate of long-term government bonds, but it presents a useful mid-point between the lower social discount rate and the higher private discount rate.

7.1.3 Case Studies: Provinces and Data on Land-use Activities

This study focuses on the two provinces of Riau and Papua, which vary in terms of their deforestation rates and causes of deforestation. As discussed in Chapter 5, Riau currently has the highest rate of deforestation, while Papua is the lowest in Indonesia. In Riau, the two major causes of deforestation are the expansion of oil palm plantations and timber plantations to supply the pulp and paper industry (Uryu et al., 2008). In contrast, pressures on Papua's forest mostly come from logging activities that cause forest degradation (Andrianto et al., 2008; Tropenbos International, 2010).

Commercial logging extraction is considered the most lucrative activity in the forestry sector (Barr et al., 2006). Public revenues collected by the national government from commercial logging comes from licence fees, forest rents, the reforestation fund, the land and building tax and personal and corporate income taxes. All public revenues generated from commercial logging extraction, except the corporate tax, are distributed amongst governmental levels using a revenue-sharing mechanism. A licence fee is paid when a timber concession is issued or renewed. The magnitude of the fees charged depends on the size of the concession area and whether it is a new licence or a renewal. The forest rent is a timber royalty, which is collected on the basis of volume and type of species harvested. The reforestation fund is non-refundable and is based on the type of species, grade and location of the wood harvested. Revenues are allocated across governmental levels according to percentages established in Law 33/2004 and Government Regulation 55/2005. The national government retains the corporate tax entirely.

If forests are converted to crop plantation activities, the government also obtains revenues from other taxes and charges. Taxes applied to plantation activities are the land and building transfer fee, the land and building tax, the personal and corporate income taxes, and the value-added tax on plantation products. The value added tax on plantation products and the corporate income tax are retained entirely by the national government. Similar to logging activities, the land and building tax and fees from plantation activities are distributed back to districts and provinces sustaining the activities within their administrative boundaries using the revenue-sharing scheme. Government regulation 48/1997 stipulates that the tax on plantations of 25 hectares or more is up to 40 per cent of the value of the land at the selling point in the market. Other revenue sources for local governments related to plantation activities are local 'fees' charged by local governments on agricultural products. The form and size of local fees on oil palm products varies between localities. A few districts in Riau charge a 'third party contribution' on the oil palm product, particularly on the fresh fruit bunches (FFB). Other districts also apply user charges on the transportation of agricultural products within their administrative boundaries.

The opportunity costs of REDD+ in this study are estimated by using data from companies currently operating in Riau and Papua provinces. The analysis focuses on three major land use activities: commercial logging operations, timber plantations and oil palm plantations. First, the net benefits derived by companies from land-use activities are estimated using a Net Present Value (NPV) analysis. Then, the forgone earnings per hectare of land accruing to the governments at all levels, including national, provincial and district governments, are estimated. The opportunity costs for all stakeholders are then converted from \$ per ha to \$ per tCO_2eq. Due to the paucity of carbon stock data for Riau and Papua provinces, the analysis uses data on the time-averaged carbon stocks of different land-use activities in Indonesia that are available in the literature.

Financial data for logging and timber plantations were derived from the working plans of five commercial logging operations in Papua and five timber plantations in Riau. Every company that is granted a licence to operate in state forests for commercial logging or timber plantation is required to submit a working plan document at the beginning of its operation. The document usually includes the harvesting plan, information on expected timber yields and a financial analysis. Data on commercial logging activities are derived from companies in Papua because only one active logging company is currently operating in Riau, where the remaining commercial timber stock is limited. Data on timber plantations are from Riau only, due to the lack of information on timber plantation companies operating in Papua. According to Indonesia's regulatory framework, only unproductive forests can be converted to timber plantations (Kartodihardjo and Supriono, 2000; Pirard, 2008). This arrangement therefore forbids timber plantations from legally opening intact primary forests. We recalculate the NPV for all companies using standard financial assumptions – particularly for the discount rate – and the timber prices obtained from the average of timber prices used by all companies.

For oil palm plantations, this study draws on data from Butler et al. (2009) and Rotheli (2007). The model presented by Butler et al. (2009) is recalculated to:

1. apply the discount rates discussed above;
2. adjust financial and yield assumptions, including oil extraction and palm kernel ratios and crude palm oil price;
3. estimate public revenues from taxes and fees for national, provincial and district governments based on the existing regulatory framework (Table 7.1);
4. carry out sensitivity analysis for changes in the price of palm oil.

In June 2011, the price was reported at US$800 per ton (which is used to derive the results presented in Tables 7.2 and 7.3)[4] and it is assumed that it would remain constant for the following 30 years. On the basis of the World Bank's price forecasts, Butler et al. (2009) developed: (i) a high-constant price scenario of US$749 per ton from 2009 to 2039, and (ii) a low-variable price scenario with a price starting at US$643 per ton in 2010 and decreasing to US$510 per ton by 2020, to then remain at that level until 2039. We report sensitivity analysis by using a low price for palm oil of US$680 per ton (which is the average of the prices used by four companies in Indonesia in 2010) and a high price of US$1,000 per ton.

Logging incomes estimated by Butler et al. (2009) are also recalculated to accommodate different conditions of the forest cleared at the beginning of oil palm plantations. Natural forests cleared for oil palm plantation are assumed to have a timber potential of 16 m^3/ha in logged-over forests and 90 m^3/ha in intact primary forests respectively. This assumption is based on the actual timber harvested by a timber plantation during land clearing at the beginning of operations in Riau.

Table 7.1 Tax and fee rates related to logging and plantations

Taxes and fees	Rate
Oil palm plantations	
Land and building tax	0.5 × 40% × determined value of the land
Value added tax	10% × price of crude oil and Kernel oil × yield
Corporate income tax	30% × net profit of companies
Land and building transfer fee	5% × determined value of the land
Commercial logging and timber plantations	
Reforestation levy	
– Dipterocarpaceae	US\$16/m^3 (Riau); US\$ 13/m^3 (Papua)
– mixed tropical hardwood	US\$13/m^3 (Riau); US\$ 10.5/m^3 (Papua)
– superior (prime) species	US\$18/m^3 (all regions)
Forest licence fee	US\$0.289 × total area (timber plantation in Sumatra)
	US\$1.67 × total area (commercial logging in Papua)
Forest resource rent	10% × volume of timber harvested × timber price
– Dipterocarpaceae	US\$7.1 × volume of timber harvested (Riau)
	US\$6 × volume of timber harvested (Papua)
– mixed tropical hardwood	US\$4 × volume of timber harvested (Riau)
	US\$2.94 × volume of timber harvested (Papua)
– superior (prime) species	US\$10.05 × volume (all regions)
– acacia	US\$0.31 × volume (all regions)
Land and building tax	0.5 × 40% × value of object at the selling point
Corporate income tax	30% × net profit of companies

Source: Government Regulation No. 59/1998; Government Regulation No. 74/1998; Government Regulation No. 92/1999; Ministerial Decree No. 859/Kpts-II/1999.

The assessment of opportunity costs accruing to governments is based on taxes and fees collected from each land-use activity (Table 7.1). As the precise rate of taxes and fees depends on the price of land and products, the average rates used in the working plan documents are used to calculate the opportunity cost faced by government. Personal income and export taxes are excluded in the calculation of the opportunity cost of all land-use scenarios. The exclusion of the personal income tax is due to data paucity related to the amount of labour required for each land-use activity. The export tax is ignored because the analysis focuses on the farm gate value.

7.1.4 Carbon Stock Data

Carbon stock data reported by Palm et al. (2004; 1999) are used as they are the most comprehensive time-averaged data for all types of land uses in Indonesia. Those authors do not, however, provide information on carbon stocks in peat swamp forest. Venter et al. (2009) provide the estimate of carbon emissions in peat and mineral forests related to forest conversion to oil palm plantation, but not for other land-use activities. In order to provide a comparison of opportunity costs in peat versus mineral forests, sensitivity analysis is conducted using carbon stock data from Venter et al. (2009).

Potential total carbon losses are converted from ton of carbon per hectare (tC/ha) to metric ton of carbon dioxide equivalent per hectare (tCO_2eq/ha) by multiplying by the molecular weight conversion factor of 3.66 (IPCC, 2006).

7.2 RESULTS: THE OPPORTUNITY COSTS OF REDD+ AND THEIR DISTRIBUTION

Oil palm and timber plantations generate the highest NPVs per hectare for all stakeholders compared to commercial logging (Table 7.2), which, in the case of oil palm, is some 20 times that of logging. It has already been noted by other studies that logging of primary or degraded forests often generates additional benefits for timber and oil palm plantations (Butler et al., 2009; Fisher et al., 2011; Grieg-Gran, 2008) and cannot be simply seen as the low-hanging fruit of opportunity costs for REDD+ activities. In the short term, there may be situations, however, in which logging is not going to be followed by plantations (given that the extent of the areas being logged are larger than those being planted) and the opportunity cost for REDD+ activities focused on degradation would then be that of logging.

Table 7.2 Average opportunity costs (NPV US$/ha) for private and public stakeholders (percentage allocation in brackets; 10% discount rate; palm oil price US$800/t)

Alternative land-use activities	Company	Government total	National	Provincial	Producing district	Other districts
Commercial logging	206 (46.68)	235 (53.32)	140 (31.71)	6 (1.29)	69 (15.58)	21 (4.74)
Timber plantation without prior logging	1,037 (64.62)	568 (35.38)	536 (33.41)	7 (0.44)	14 (0.90)	10 (0.63)
Timber plantation with prior logging in degraded forests	1,507 (58.75)	1,058 (41.25)	767 (29.92)	29 (1.14)	213 (8.29)	49 (1.90)
Oil palm plantation without prior logging	6,355 (57.97)	4,608 (42.03)	4,587 (41.85)	3 (0.03)	17 (0.15)	0 (0)
Oil palm plantation with prior logging in degraded forests	6,458 (57.45)	4,782 (42.55)	4,678 (41.62)	10 (0.09)	82 (0.73)	13 (0.11)
Oil palm plantation with prior logging in primary forests	7,099 (56.34)	5,502 (43.66)	5,057 (40.13)	34 (0.27)	350 (2.78)	61 (0.48)

In addition to the benefits obtained by companies, which have normally been considered by studies of the opportunity costs of REDD+ (for example, Butler et al., 2009; Venter et al., 2009) the various levels of government also receive revenues from taxes and fees applied to land-use activities. Under all scenarios of discount rates, companies obtain a higher portion of the NPV than the government from the land-use activities considered, with the exception of commercial logging (Table 7.2).

The central government retains the largest shares of revenues from land-use activities compared with local governments. In the case of logging, the producing districts derive about half as much as the central government. The benefits derived by the districts from commercial logging mostly come from the reforestation tax, given that 40 per cent of those revenues collected by the national government are returned to the producing regions. For the other land-use activities, local governments retain a very small share of the revenues. Although the portion of benefits received by district governments is meagre in comparison to the national level, in absolute terms they derive more from oil palm and timber plantations than from logging. Therefore, they too have a financial incentive to seek the conversion of forests to plantations.

To assess whether the opportunity costs of REDD+ are competitive with the costs faced by developed countries to reduce emissions domestically, we consider the opportunity cost of carbon emissions. We compare the opportunity costs of REDD+ current carbon prices because the idea underlying the establishment of REDD+ is that it is supposed to be an early and relatively cheap way of reducing carbon emissions (Stern, 2006).

If only companies were to be compensated, it would be appropriate to consider the costs resulting from the application of either the 10 or the 15 per cent discount rates. At these rates, the opportunity cost of REDD+ for all land uses, except for oil palm plantations on mineral soil with logging in degraded forest (Table 7.3), was competitive with the price of carbon on the European market (EUA), which is in the range of US$16–17 at the time this analysis was undertaken.[5] The establishment of oil palm plantations on peat soils generates considerably larger carbon emissions than the other oil palm plantation types and presents, therefore, significantly lower opportunity costs.

From the government's perspective, the case of the 5 per cent

Table 7.3 Minimum REDD payments to offset opportunity costs (US$/ton$CO_2$eq; palm oil price US$800/t)

Land-use activities	Carbon loss[#] (tCO_2/ha)	Company	Government total	National gov	Provincial gov	Producing district	Other districts
		15% discount rate					
Commercial logging in primary forest	779.58	0.18	0.21	0.12	0.01	0.06	0.02
Timber plantation with prior logging in degraded forest	135.42	6.59	5.14	3.33	0.17	1.34	0.29
Oil palm plantation:							
– with prior logging in degraded forest on mineral soil	197.64	13.84	12.52	12.08	0.04	0.34	0.05
– with prior logging in primary forest on mineral soil	977.22	3.37	3.21	2.84	0.03	0.30	0.05
– with prior logging in primary forest on peat soil	2,249	1.34	1.37	1.20	0.01	0.13	0.02
		10% discount rate					
Commercial logging in primary forest	779.58	0.26	0.30	0.18	0.01	0.09	0.03
Timber plantation with prior logging in degraded forest	135.42	11.13	7.81	5.67	0.22	1.57	0.36
Oil palm plantation:							
– with prior logging in degraded forest on mineral soil	197.64	32.67	24.20	23.67	0.05	0.41	0.06

– with prior logging in primary forest on mineral soil	977.22	7.26	5.63	5.17	0.03	0.36	0.06
– with prior logging in primary forest on peat soil	2,249	2.95	2.39	2.20	0.02	0.16	0.03
		5% discount rate					
Commercial logging in primary forest	779.58	0.45	0.50	0.30	0.01	0.14	0.04
Timber plantation with prior logging in degraded forest	135.42	21.79	12.06	9.40	0.30	1.89	0.48
Oil palm plantation:							
– with prior logging in degraded forest on mineral soil	197.64	77.55	50.31	49.64	0.06	0.52	0.08
– with prior logging in primary forest on mineral soil	977.22	16.45	11.09	10.52	0.04	0.45	0.08
– with prior logging in primary forest on peat soil	2,249	6.76	4.71	4.46	0.02	0.19	0.03

Notes:
#: Emissions from primary and degraded forest on mineral soil based on Palm *et al.* (2004).
Emissions from forest on peat soil based on Venter *et al.* (2009).

discount rate may be more appropriate, however, for the reasons discussed earlier. In this scenario, commercial logging and oil palm plantations on peat soil with prior logging in primary forest (which is not supposed to take place according to the regulatory framework) have a total break-even cost well below current carbon prices (Table 7.3). Oil palm plantations on mineral soil with prior logging in primary forest (which is also not supposed to take place) have a total break-even cost similar to carbon prices noted earlier. The other two land uses present prohibitive costs (Table 7.3). It should be noted that while the opportunity cost of logging is competitive with carbon prices, it is the first commercial activity that can take place in primary forests and, after years of logging, when forests become degraded, timber plantations are allowed, according to existing regulations, to clear-cut and replace the degraded forest. Therefore, the opportunity cost of commercial logging cannot be considered in isolation.

The sensitivity analysis (Table 7.4) shows that a lower price for palm oil would not make the opportunity cost of oil palm plantations in degraded forest competitive with carbon prices if the government chose to use the 5 per cent discount rate. At the other extreme, oil palm plantations in peat soil present a break-even price for carbon that is lower than carbon prices – even with a higher palm oil price and a 5 per cent discount rate.

7.3 DISCUSSION

Previous estimates of the minimum REDD+ payment for oil palm plantations in Indonesia were in the range of US$9.85–33.44 per tCO_2 in mineral soil forests and US$1.63–4.66 per tCO_2 in peat areas (Venter et al., 2009) and US$7.66–19.24 per tCO_2 (Butler et al. 2009; without reference to soil types). Venter et al. (2009) assumed a carbon stock lower than other studies (for example, Butler et al., 2009; Palm et al., 2004; Palm et al., 1999) resulting in a higher estimate of the opportunity costs.

We demonstrate that, at palm oil prices reported in Table 7.4, using the higher carbon stocks reported in the literature, and a mid-range discount rate of 10 per cent, the minimum REDD+ payment to compensate for the opportunity cost amounts to about US$57 per tCO_2 for the case of plantations on mineral soils and logging in degraded forest (the legal option). This estimate is well above those previously

Table 7.4 Sensitivity of minimum REDD+ payments to changes in the price of palm oil (USD per tonCO_2eq)

Discount rate:	5%		10%		15%	
Land-use activity and price of palm oil per ton	Company	Government total	Company	Government total	Company	Government total
Plantation with prior logging in degraded forest: mineral soil						
– US$ 680	56.56	40.51	22.49	19.53	8.38	10.08
– US$ 800	77.55	50.31	32.67	24.20	13.84	12.52
– US$ 1,000	112.54	66.64	49.65	31.97	22.94	16.59
Plantation with prior logging in primary forest: mineral soil						
– US$ 680	12.23	9.11	5.22	4.69	2.27	2.72
– US$ 800	16.45	11.09	7.26	5.63	3.37	3.21
– US$ 1,000	23.49	14.39	10.67	7.20	5.18	4.04
Plantation with prior logging in primary forest: peat soil						
– US$ 680	4.93	3.85	1.56	1.98	0.86	1.15
– US$ 800	6.76	4.71	2.95	2.39	1.34	1.37
– US$ 1,000	9.82	6.14	4.43	3.08	2.13	1.73

reported and, like the other estimates, does not include other costs, such as the management of REDD+ activities and transaction costs. Given that large areas of forest have been degraded −55.6 per cent and 49.4 per cent of secondary forests in areas classified as production forests and in all areas classified as forests respectively (Ministry of Forestry, 2008b) – it seems that the establishment of REDD+ activities in those areas may be too costly if the development of oil palm plantations is an option.

The good news from this analysis is that all the other land-use activities (including oil palm plantations on peat soil) present minimum REDD+ payments that are competitive with carbon prices at the discount rates of 10 and 15 per cent. The minimum REDD+ payment to offset the opportunity costs of oil palm plantations on peat soil is always rather cheap, including the case of a 5 per cent discount rate.

In relation to the distribution of the revenues for the different land-use activities, the foregoing analysis shows that the total revenues derived by the various government levels is very large and, in the case of commercial logging, exceeds that retained by the companies (Table 7.2). The national government obtains a very large share of the benefits, so it has a strong interest in promoting all types of land-use change. Going below the surface, each of the national level ministries (sectors) has different interests in the pursuit of the alternative land-use activities. The Ministry of Forestry has an interest in commercial logging in (natural) production forests to generate the lucrative reforestation levy. The Ministry of Forestry retains as much as 60 per cent of the fund and controls its utilisation to support nationally based forestry programmes and policies (Barr et al., 2009). In contrast, the Ministry of Agriculture, which is responsible for the promotion of agricultural development, promotes oil palm plantations and/or other agricultural activities in conversion forests, which total 22 million hectares. Revenues generated from oil palm plantations – which have reached 5.2 million hectares (Ministry of Agriculture, 2009 quoted by Rist et al., 2010) – mostly from the value added tax and the corporate income tax, contribute to the national government's general income to finance a wide range of services.

As for local level governments, they receive a very small share of the revenues from the various land uses. However, they (particularly the producing districts) do receive benefits from timber and oil palm plantations. This explains the fact that forest agencies at the district level are interested in promoting timber plantations like their counterparts at the national level. The permits for commercial logging and timber plantations are issued by the Ministry of Forestry, although district governments can submit a proposal for activities at the district level (Resosudarmo et al., 2006). Local governments' support is also crucial for licences issued at the national level, as strong resistance from local stakeholders can hinder companies' operations, as reported in a number of regions in Indonesia.[6]

Local governments are also interested in expanding oil palm plantations in their localities (McCarthy and Cramb, 2009; Rist et al., 2010; Sandker et al., 2007; Zen et al., 2005). In the case of oil palm plantation, local governments have the authority to issue a business permit, which is required before the final decision on forest clearance can be made by the Ministry of Forestry (Colchester et al., 2006). Therefore, district governments have more influence in this decision-making process compared with commercial logging and timber plantations. Although district governments obtain a small portion from the total benefits of oil palm, some regions apply a local fee, which is called a third-party contribution, to oil palm products.

The interests of district governments to support proposals for natural forest conversion to productive land-use activities could also be due to other economic and political benefits. Heads of districts might have a greater chance of maintaining their popularity if they attract investments and generate local revenues. Oil palm plantations, for instance, also generate employment and livelihoods for local people (McCarthy and Cramb, 2009; Rist et al., 2010; Sandker et al., 2007; Susila, 2004; Zen et al., 2005). Susila (2004) claims that oil palm activities can contribute as much as 63 per cent of smallholder household incomes in two locations in Sumatra. Sandker et al. (2007) simulated the impact of conversion of forests to oil palm in Malinau district (Kalimantan) and found that the total number of formal jobs created by oil palm development (22,000–120,000) could exceed the employment opportunities generated by mining, logging, and the civil service put together (10,000). Moreover, Rist et al. (2010) reported that local elites obtained financial support during electoral campaigns by providing their support for the establishment of oil palm plantations.

7.4 CONCLUSION

Research on REDD+ casts doubt on its financial viability, given the seemingly high opportunity costs. This chapter shows that REDD+ might not be able to compete with some alternative land uses associated with very high opportunity costs, such as oil palm plantation in degraded forests on mineral soils. However, we also demonstrate that in some cases REDD+ activities may be a viable option to reduce deforestation and degradation. Oil palm and timber plantation companies are keen to operate in natural forests so that they can obtain additional logging income prior to their operations. When converted further to dollars per tCO_2, the additional benefit from logging income is very low, as logging natural forests results in very high emissions. Therefore, REDD+ payments could, for instance, be allocated to compensate stakeholders for relocating proposed oil palm plantation expansion from natural forests to non-forested areas or degraded forests. REDD+ would not halt oil palm expansion, but would help incentivise stakeholders to keep natural forests intact. Furthermore, REDD+ payments could also prevent the conversion of peat forests, as it results in extremely high emissions and it is therefore associated with low opportunity costs.

Previous studies of the opportunity costs of REDD+ have not sufficiently considered the influence of the discount rate on their results. This study demonstrates that the competitiveness of some avoided deforestation activities in terms of the opportunity cost of reduced emissions can be significantly affected by the choice of the discount rate. Both governments and those conducting further studies of the opportunity costs of REDD+ should therefore consider which discount rate might be most appropriate for their specific conditions, and carry out detailed sensitivity analysis.

From this chapter, it is also evident that the design of the distribution of revenues generated from REDD+ activities should consider the specific incentives of the different government levels and across sectors, which have specific interests in promoting certain land-use changes. Without proper compensation, subnational governments will have no incentives to support REDD+, as they would face forgone taxes, fees and shared-revenues, while the benefits of REDD+ reach far beyond their administrative boundaries (Ring et al., 2010). REDD+ payments, at the very minimum, should compensate the costs of REDD+ implementation, including opportunity costs (Boucher, 2008; Cattaneo, 2008; Strassburg et al., 2009). However, REDD+ payments based simply on the opportunity costs incurred by district governments might not be sufficient to change local governments' behaviour. District governments also see other benefits from the support of land-use change activities, such as job creation and other institutional benefits as mentioned before. Therefore, REDD+ schemes should provide a revenue stream that is higher than the alternative land-use activities in order to demonstrate that it is a superior option to the latter activities. The relevant local government levels could therefore use those REDD+ payments to provide improved services to their citizens.

In order to determine the size of IFTs to be provided to local governments for REDD+ revenue distribution, different approaches of the distribution formula need to be tested, and are discussed in the next chapter.

NOTES

1. This chapter is a slightly edited version of Irawan, S., Tacconi, L. and Ring, I. (2013) Stakeholders' Incentives for Land Use Change and REDD+: The Case of Indonesia. *Ecological Economics*, 87, 75–83.
2. Butler et al. (2009) report REDD+ opportunity costs from oil palm plantations are US$3,835–9,630 per hectare, while the net carbon saving of avoided conversion is 149 tons/ha.
3. http://www.tradingeconomics.com/Economics/Government-Bond-Yield.aspx?symbol=IDR, accessed 18 August 2011.
4. The Economist Intelligence Unit, http://gfs.eiu.com/Article.aspx?articleType=cfs&articleId=1688295353, accessed 24 June 2011.
5. http://www.pointcarbon.com, accessed 5 June 2011.
6. http://www.riaumandiri.net/rm/index.php?option5com_content&view5article&id514559:datangi-kantor-bupati-meranti-ratusan-massa-demo-tolak-hti-&catid540:riau-raya; http://dpd.go.id/2010/01/dpd-desak-menhut-hentikan-izin-usaha-hutan/, accessed 11 June 2011.

8. The distribution formulae of IFTs for REDD+[1]

Following the assessment of REDD+ opportunity costs incurred by local governments from alternative land-use activities, this chapter considers the options for the specification of the formula to determine the size of IFTs for the distribution of REDD+ revenue to district governments. Two important aspects of the distribution formula that are explored are the grant size to be allocated to different levels of government; and the size of IFTs allocated for each eligible locality to pursue REDD+. The grant size, or distributable pool, is the vertical dimension of IFTs that determines the total size of grants or transfers distributed to different levels of government; whilst, the horizontal dimension decides the size of transfers allocated to each local government unit (Bird, 1999; Bahl, 2000).

As discussed in previous chapters, IFTs for biodiversity conservation can help reconcile local costs with spillover benefits of conservation that reach beyond local boundaries. Brazil and Portugal use IFTs to support biodiversity conservation by transferring a portion of national or state governments' taxes (e.g. in Brazil, state-level value added tax) to local levels on the basis of conservation and ecological indicators (Grieg-Gran, 2000; May et al., 2002; Ring, 2008c; Santos et al., 2012). Several studies have also suggested transferring a portion of national or state governments' revenue to local levels to compensate for the management and the forgone opportunity costs borne by localities with protected areas (Köllner et al., 2002; Ring, 2008b; Kumar and Managi, 2009). To determine the size of transfers for conservation at the local level, indicators, such as conservation units, biodiversity index, and land area protected, have been proposed (e.g. Ring et al., 2010; Ring, 2008a; Ring, 2008b; Ring, 2008c; Köllner et al., 2002; Kumar and Managi, 2009; Santos et al., 2012).

Using IFTs to channel REDD+ payments to local governments requires a new approach to the determination of the size of transfers.

As discussed in Chapter 7, using opportunity costs as the basis for REDD+ revenue distribution to local governments could be problematic as they do not consider other economic benefits as well as the institutional and political benefits that may no longer be obtained if local governments choose to pursue REDD+ in their localities. Moreover, as a REDD+ scheme would involve the transfer of financial resources from developed to participating developing countries, the purpose of IFTs would not merely be to correct spatial externalities of conservation but also to distribute that revenue, which could exceed the opportunity costs, vertically between government levels. REDD+ revenue can therefore be distributed using a vertical revenue-sharing scheme, which is commonly used to distribute taxes and fees collected by the national government. The size of the vertical revenue sharing is usually based on a share of a national tax and the size of taxes collected within certain administrative boundaries (Bird, 1999; Bahl and Wallace, 2007).

Here it is assumed that REDD+ would be implemented using a nationally based implementation approach in which the national government would receive REDD+ payments and there are no direct payments from the international level to local governments (Angelsen et al., 2008). Scholars have also proposed a nested approach to the implementation of REDD+ in which subnational projects could be allowed to receive payments directly from international buyers (e.g. Pedroni et al., 2009). Busch et al. (2011) assume that the nested approach would be adopted and estimate the incentive structure required for local governments in Indonesia to participate in REDD+. That approach, however, ignores the existing political economy of land-use change in Indonesia, including the distribution of power between the central and local governments and the existing incentive structures influencing different stakeholders in the pursuit of forest exploitation and land-use change. The national government currently retains most of the revenue from land-use alternatives to REDD+ as discussed in Chapter 7. This situation is also common in other decentralised countries, where the higher level of government collects the largest share of public revenue and distributes part of it to local governments using IFTs (de Mello, 2000). Hence, IFTs are still a cornerstone of subnational government financing in most developing and transition countries (Bahl, 2000).

The chapter proceeds by first simulating the reference emissions level (REL) in Riau and Papua. RELs determine the business-as-usual scenario of carbon emissions that would be released from deforestation and forest degradation in the future without REDD+ (Meridian Institute, 2009). Different possible approaches to calculate RELs result in different amounts of emission reductions achieved by a locality, which in turn would affect the size of REDD+ incentives to be allocated to the sub-national level (Cattaneo, 2011; Busch et al., 2009). The methods adopted to estimate the grant size for different government levels and the size of IFTs for eligible district governments are then discussed. After presenting the results of the analysis, the chapter discusses the advantages and disadvantages of the cost-reimbursement and the derivation approaches for the design of the IFTs and the implications of the findings for designing REDD+ payment distribution.

8.1 DETERMINING REFERENCE EMISSION LEVELS (RELs) IN RIAU AND PAPUA

Chapter 5 discussed the role of local governments in forest management and the deforestation rates in Riau and Papua. This section summarises the historical deforestation rates and the future land-use change that may take place based on the spatial plan and forest classifications in Riau and Papua. Based on Forestry Law 41/1999, state forestlands in Indonesia are classified into production, protection and conservation forests. The main function of production forests is to produce forest commodities, mainly timber. Some production forests are also classified as conversion forests, which can be legally converted to other non-forest land-use activities. Exploitation activities, however, cannot take place in protected and conservation forests, as the law stipulates that those forests are intended to provide environmental services and to conserve biodiversity. In order to determine the RELs, this section focuses mainly on legal deforestation that can take place in the production and conversion forests that still have forest covers (Table 8.1 and Table 8.2).

The total area classified as state forestlands in Riau comprises 8.6 million hectares. Approximately 89 per cent of the total area is classified as production forest, where 50 per cent can be converted

Table 8.1 Remaining production and conversion forest area in Riau

Forest classification	Total area (ha)	Area with tree cover (ha)	Activities permitted by law Annual deforestation (ha/yr)
Production forest land without active concessions	712,614	712,614	Timber plantation – actual forest cover change to timber plantation was 37,943 ha/year between 1982 and 2007 (Uryu *et al.*, 2008) – timber plantation licences issued were for 106,625 ha/year (The Ministry of Forestry, 2008c)
Production forest land under commercial logging concessions	1,207,003	804,692	
Conversion forest land	4,107,500	620,100	Oil palm plantation – forest cover change to oil palm plantation was 44,000 ha/year between 1982 and 2007 (Uryu *et al.*, 2008)
Total	6,027,117	2,137,406	

to other land-use activities and the remaining 39 per cent is classified as permanent production forest. Both commercial logging and timber plantations can operate in permanent production forests. However, as commercial logging is no longer feasible in Riau, the expansion of timber plantations replaces the logged-over production forests. The issue of licences for timber plantations in Riau was approximately 106,625 hectares/year between 2001 and 2008, although the actual conversion of forests to acacia plantations was reported at 37,943 hectares/year between 1982 and 2007 (Ministry of Forestry, 2008a).[2] Forests that are currently not under concession are mostly secondary forests with no potential for commercial logging, and can be converted to timber plantations. The total production forest area, with and without active licences, that could be further converted to timber plantations amounts to 1,517,306 hectares (Ministry of Forestry, 2008a). Assuming the continuation of the current annual land-use change, all the remaining production forests would be converted to timber plantations over the next 40 years (Table 8.1).

Another major cause of deforestation is the expansion of oil palm plantations, which can take place in conversion forests. The total forest loss, due to oil palm plantations, was estimated at

1,113,090 hectares (around 44,000 hectares/year) between 1982 and 2007 (Uryu et al., 2008). In Riau, the remaining area of conversion forests with tree cover was estimated at 620,100 hectares (The Ministry of Forestry, 2008c) (Table 8.1). Should the existing annual land-use change for oil palm plantations persist, Riau is expected to lose all the remaining forest cover in areas designated as conversion forests during the next 14 years.

In contrast to Riau province, Papua is home to the largest area of remaining tropical forests in Indonesia, with the lowest rate of deforestation in the country. In Papua, around 31 million hectares are designated as forest zone, of which 10 million hectares are classified as production forests and an additional 6 million hectares as conversion forests. Moreover, approximately 14 million hectares are protected as conservation and protection forests. The main productive activity taking place in Papua's forests is commercial logging. If the current trend continues in Papua province, all the remaining production forests currently without active concessions would be allocated to logging concessionaires over the next 30 years (Table 8.2). Commercial logging usually results in severe forest degradation. The areas degraded and deforested within the commercial logging concessions between 2000 and 2005 were reported at 709,968 hectares and 71,666 hectares respectively (Andrianto et al., 2008). At the existing annual rate of degradation (65,231 hectares/year), 1.96 million hectares of forests currently under logging concessions would be degraded over the next 30 years. Furthermore, between 1992 and 2008, the issue of licences to convert forests to oil palm plantations in Papua reached 318,550 hectares, or around 19,909 hectares annually (Ministry of Forestry, 2008c). Land-use change from forests to crop plantations in Papua and West Papua was reported at 7,510 hectares and 25,201 hectares between 2000 and 2005 and 2005 and 2008 respectively (Tropenbos International, 2010). With the current annual rate of expansion of oil palm plantations, which is 2,520 hectares in Papua province alone, deforestation would amount to 75,000 hectares over the next 30 years.

Table 8.2 Remaining production and conversion forest area in Papua

	Total area (ha)	Area with tree cover (ha)	Activities permitted by law Annual deforestation/degradation (ha/yr)
Production forest land without active concessions	3,845,902	3,845,902	Commercial logging – logging licences issued between 1988 and 2008 in Papua were for 300,000 ha/year
Production forest land under commercial logging concessions	6,173,980	4,386,857	Total degradation in logging concession between 2000–2005 – 709,968 ha in both Papua and West Papua provinces (Andrianto *et al.*, 2008) or 404,687 ha in Papua province alone, which equates to 80,935 ha/year – 572,208 ha in both Papua and West Papua provinces (Tropenbos International, 2010) or 326,158 ha in Papua alone, which equates to 65,231 ha/year
Conversion forest land	6,568,816	4,795,236	Crop plantation – plantation licences issued between 1992 and 2008 cover 318,550 ha in Papua province alone or around 19,909 ha/year – forest cover change to crop plantation in both Papua and West Papua provinces was 7,510 ha between 2000 and 2005 and 25,201 ha between 2005 and 2008 (Tropenbos International, 2010). Annual expansion of oil palm plantation is around 12,601 ha/year
Total	16,588,698	13,027,995	

8.2 COMPARING DIFFERENT APPROACHES TO DETERMINE REFERENCE EMISSION LEVELS (RELs)

RELs for Riau and Papua may be developed using the different approaches being proposed in the literature at the global level, including:

1. average emissions from deforestation over a recent historical reference period (Santilli et al., 2005);
2. a combined incentives mechanism, which combines higher reference emission levels for countries with historically low deforestation rates and lower reference emission levels for countries with historically high deforestation rates (Strassburg et al., 2009);
3. the annualised fraction of the volume of terrestrial carbon stocks estimated to be at risk of emission in the long run, based on biophysical, economic and legal considerations (Ashton et al., 2008).

When determining a REL for a locality it is important to consider the remaining carbon stocks in standing forests and alternative land-use activities. A locality with a historically high deforestation rate, such as Riau, normally only has a small amount of carbon stocks stored in the remaining standing forests. Thus, even if the historical deforestation rate persists, the trend of carbon emissions from deforestation and forest degradation is declining due to diminishing carbon stocks in standing forests. The approach to determine RELs proposed by Ashton et al. (2008) considers the remaining forest areas that are at risk. However, this approach ignores the existing practice of deforestation and uses only time as the indicator to estimate the total area of forests that will be deforested. Furthermore, in the case of Indonesia, certain alternative land-use activities can only take place legally in a particular forest classification. We assume that forest classifications would not be changed unless the existing regulations were amended. Hence, estimation of the RELs needs to examine both the historical deforestation rates caused by particular land-use activities as well as the remaining forests in forest classifications that could be converted to the land uses in question.

This chapter examines another approach to establishing the REL by combining three important indicators: (i) the historical deforestation rate at the local level; (ii) the remaining carbon stocks in standing forests that can be legally converted; and (iii) the alternative land-use activities that can legally take place in particular forest classifications. The following formulae formally describe the four approaches to calculating reference emission levels simulated in the chapter.

1. Approach 1: historical reference emission levels (Santilli et al., 2005 cited in Busch et al., 2009)

$$Bi = Hi$$

where,
Bi = reference emission level (baseline) for locality i (tCO2eq)
Hi = historical emission level (business as usual) for locality i (tCO2eq)

2. Approach 2: reference emission level is weighted average of national and local historical rates (Strassburg et al., 2009 cited in Busch et al., 2009)

$$Bi = [\alpha \times Di + (1 - \alpha) \times GAD] \times CDi \times 3.67$$

where,
Di = historical deforestation rate for locality i (ha/yr)
α = weight placed on the national historical deforestation rate
GAD = national average deforestation rate (ha/yr)
CDi = carbon density for locality i (tC/ha)
3.67 is the atomic ratio of carbon dioxide to carbon

3. Approach 3: the reference emission level is the annualised fraction of forest carbon at risk of emission (Ashton et al., 2008 cited in Busch et al., 2009)

$$Bi = Ai/T$$

where,
Ai = forest carbon stocks at risk of deforestation over the long term in locality i (tCO2eq)
T = time over which forest carbon stocks is at risk (years)

4. Approach 4: reference emission level is based on a combination of the historical deforestation rate and the remaining carbon stocks in forests that can be legally converted to other land-use activities.

$$Bi = (Fi \, ' \, Gi)/Di$$

where,
Fi = areas that can be converted to other land-use activities in locality i (ha)
Gi = carbon emissions released due to land-use change from forest to other land-use activities (or emission factors) (tCO2eq/ha)
Di = historical deforestation rate for locality i (ha/yr)

Table 8.3 Reference emission levels (RELs) using four different approaches for Riau and Papua

Approach	Annual emissions (tCO_2eq)	Emissions over 30 years (tCO_2eq)
Papua		
1	247,324,581	7,419,737,416
2	210,425,660	6,312,769,804
3	165,565,144	4,966,954,314
4	113,390,197	3,401,705,907
Riau		
1	11,096,721	332,901,632
2	9,439,922	283,197,648
3	7,447,380	223,421,391
4	9,223,460	276,703,796

Using the historical rate of deforestation (Approach 1) results in the highest estimate of carbon emissions (Table 8.3) since it does not consider the remaining forest areas that can actually be converted to other land-use activities. For instance, the annual deforestation from oil palm plantations in Riau is around 44,000 hectares. Should the existing rate persist, total deforestation over the next 30 years would amount to 1.32 million hectares. However, the remaining forests that legally can be converted to oil palm plantations in Riau amount to only 620,100 hectares. Thus, carbon emissions from deforestation and forest degradation could be expected to decrease due to the diminishing forest stocks. Approaches 3 and 4 consider the remaining forests that can be legally converted to other land-use activities. As previously mentioned, we assume that the present forest classifications stipulated by regulations would not be amended.

Approach 2 assigns weights for both the historical deforestation rate and the carbon stocks in standing forests (Strassburg et al., 2009). The assumptions used in this study are similar to those of Busch et al. (2009), where the national average deforestation rate for Indonesia is 0.47 per cent, while the weights assigned to the historical deforestation rate and to the total carbon stocks in standing forests are 0.85 and 0.15 respectively. Assigning a weight to the total carbon stocks in standing forests within the administrative boundaries results in lower total carbon emissions compared with Approach 1.

Approach 3 assumes that the carbon stocks in forests at risk are emitted during the next 50 years (similar to the assumption made by Busch et al., 2009). Under this assumption, all forest areas that are currently classified as conversion forests in Papua (totalling 2.87 million hectares) will be converted to oil palm plantations over the next 30 years. In contrast, Approach 4 assumes that carbon stocks in standing forests at risk (or forests that can be legally converted to other land uses) will be deforested based on the historical rate. Using this approach, the total deforestation caused by oil palm plantations in Papua would only be 378,015 hectares over the next 30 years.

Applying different approaches results in varying estimates of RELs, which will eventually determine the assessment of performance achieved by each locality in reducing carbon emissions. Using the historical rate approach would result in an overestimate of the reduction in total emissions. Thus, we argue that the remaining forests that can be converted to other land-use activities should be considered in the estimation of REL. Although Approach 3 considers the remaining forests (forests at risk), it ignores the historical deforestation rate. In this chapter, therefore, we use Approach 4 to generate the RELs on which we then make a simulation of the IFT size to distribute REDD+ revenues, which considers both the remaining forests (carbon stocks) within the administrative boundaries and the historical deforestation rate.

Following estimation of the RELs, the IFT size allocated to Riau and Papua is simulated. It was assumed that all localities in Indonesia had reduced emissions by 50 per cent from the business-as-usual (BAU) level in the next 30 years. This figure is set in proportion to Indonesia's commitment to reduce emissions by 26 per cent (without international support) or 41 per cent (with international financial support) from the BAU level by 2020 as stipulated in Presidential Decree 61/2011. In order to determine the RELs and total carbon emissions reduced at the district level, the average deforestation rate of all districts is assumed to be similar to that at the provincial level. Hence, the amount of carbon emissions released by each district will vary depending on the total production forests that can be legally converted to other land-use activities (Figure 8.1). Data on forests classified as conversion forests at the district level in Riau were unavailable at the time of the analysis. For this reason, the fiscal simulation analysis for districts in Riau focuses on production forests with and without active concessions.

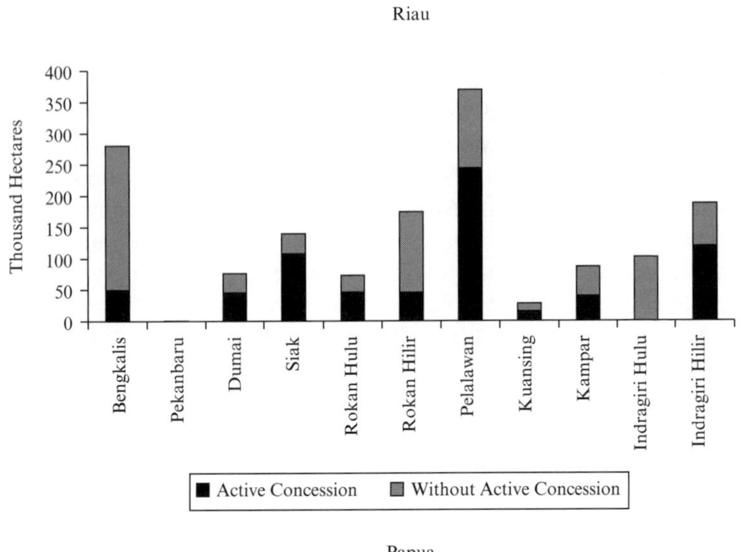

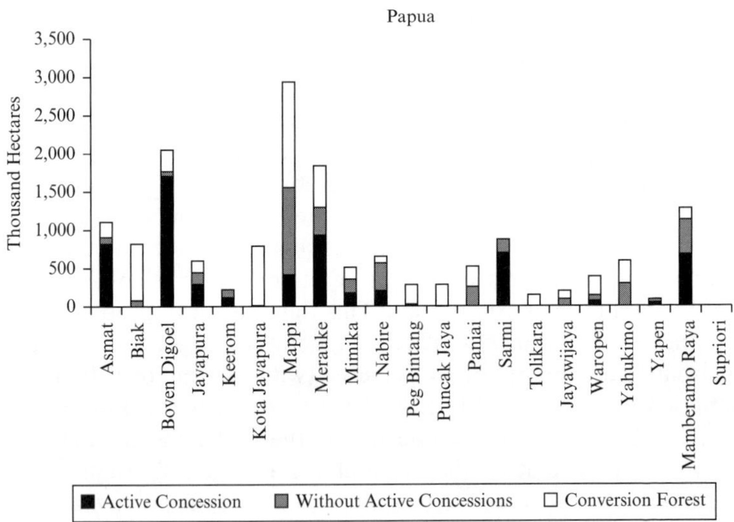

Source: Ministry of Forestry, 2008c.

Figure 8.1 Status of production forest in the districts of Riau and Papua (2008)

8.3 DISTRIBUTION FORMULAE OF IFTs FOR REDD+ IN INDONESIA

This study assumes that the central government would obtain a portion of REDD+ payments through taxes and fees collected from REDD+ projects implemented by private companies. The involvement of the private sector in REDD+ should be encouraged in forest areas that are currently under active concessions. On the other hand, the benefits generated by REDD+ projects operating in forest areas without active concessions could be retained entirely by the state, since private companies do not need to be compensated. To determine the size of REDD+ revenues to be distributed to local government levels, the derivation approach (Option 1) and the (opportunity) cost-reimbursement approach (Option 2) are simulated.

The derivation approach (Option 1) determines the total grant size and the magnitude of IFTs for each district government based on a specified percentage of the total taxes or fees collected from REDD+ within a locality. The calculation of the grant size and the magnitude of IFTs allocated to eligible district governments under Option 1 uses the following formula:

$$\mathrm{IFT}_i = \alpha_\mathrm{d} \times \mathrm{P} \times \mathrm{Q}_i$$

$$\mathrm{GSIFT}_\mathrm{D} = \sum_{\mathrm{i}=1}^{\mathrm{z}} \mathrm{IFT}_i$$

Where,
IFT_i is the magnitude of IFT for district i
α_d is the percentage of revenues distributed to the district level
P is the price ($/$tCO_2$) of carbon
Q_i is the total carbon emission reduction in district i (tCO_2e)
$\mathrm{GSIFT}_\mathrm{D}$ is the grant size for the district level using the derivation approach.

To decide the percentage of revenues distributed to the district level, this study refers to Ministerial Decree 36/2009 on REDD+ Implementation (Table 8.4). Ministerial Decree 36/2009 sets the portions of REDD+ revenues allocated for different stakeholders, including companies, various governmental levels and local communities. The decree has been criticised by some stakeholders, as it

Table 8.4 Revenue distribution between stakeholders (percentages)

Activity	Company	Total government	National	Provincial	Producing district	Other district	Local community
		Based on existing opportunity costs					
Commercial logging in primary forest	45.4	54.6	31.9	1.4	16.7	4.6	
Timber plantation in degraded forest	59.2	40.8	29.8	1.1	8.0	1.9	
Oil palm plantation in degraded forest	54.1	45.9	44.7	0.1	1.0	0.2	
Oil palm plantation in primary forest	55.3	44.8	39.9	0.4	3.8	0.7	
		Based on the Ministerial Decree 36/2009					
REDD+ in production forest	60.0	20.0	8.0	4.0	8.0	0.0	20.0

Source: Ministerial Decree 36/2009.

does not appear to be based on a detailed quantitative assessment of the share to be allocated to each stakeholder group. This study, however, uses the portions regulated by the decree only to simulate the impact of distributing revenues between different stakeholders using the proposed percentages. Ministerial Decree 36/2009 allocates the largest share of revenues to companies, the second largest share to the local community and the remainder is shared between government levels, with the central government receiving a significantly smaller share of the total revenues than the private and public revenues arising from current land-use activities (Table 8.4).[3] The distribution stipulated in the decree differs significantly from the existing allocation of revenues to different stakeholders from land-use alternatives to REDD+, which sees the central government receiving the largest share of revenues as discussed in Chapter 7. The Decree appears, therefore, to provide incentives to the various stakeholders (except for the central government) to choose REDD+ activities rather than land-use alternatives, as long as the absolute value of the former is not lower than the latter. The non-producing districts are the only stakeholders who would not receive a share of REDD+ revenues.

Option 2 determines the grant size and the size of IFTs to eligible units based on the cost-reimbursement approach (the costs of REDD+ at the local level). Ideally, the cost-reimbursement approach should include the estimation of all costs of REDD+, including opportunity, management and transaction costs. However, this study focuses only on the opportunity costs due to the paucity of data related to REDD+ management and transaction costs. It draws on the analysis of REDD+ opportunity costs reported in Chapter 7, which discusses the share of revenues accruing to the various stakeholders on the basis of ongoing land uses, taxation framework and allocation of revenues among the different government levels. The distribution formula for the cost-reimbursement approach is as follows:

$$IFTCRi = (OC_{doil} \cdot Qi_{oil}) + (OC_{dtimber} \cdot Qi_{timber}) + (OC_{dlogging} \cdot Qi_{logging})$$

Where,
$IFTCR_i$ is IFT for District i based on the cost-reimbursement approach

OC_{Doil} is the opportunity cost accruing to the district level from palm oil ($/t$CO_2$e)
$OC_{Dtimber}$ is the opportunity cost accruing to the district level from timber plantation ($/t$CO_2$e)
$OC_{Dlogging}$ is the opportunity cost accruing to the district level from commercial logging ($/t$CO_2$e)
Qi_{oil} is the total carbon emission reductions in District i from avoided oil palm plantations (tCO_2e)
Qi_{timber} is the total carbon emission reductions in District i from avoided timber plantations (tCO_2e)
$Qi_{logging}$ is the total carbon emission reductions in District i from avoided commercial logging (tCO_2e).

8.4 RESULTS: GRANT SIZE AND THE SIZE OF IFTs

Using the cost-reimbursement approach, the size of grants allocated to provincial and district governments will be similar to the opportunity costs incurred to reduce deforestation and forest degradation. The distribution of REDD+ revenues therefore considers the cost curve of reducing emissions in each province and district, which varies depending on the condition of the forests and land-use alternatives. If all provinces had to reduce deforestation and forest degradation by 50 per cent below the BAU level, the total emission reductions would be 138 million tCO_2eq and 1,700 million tCO_2eq in Riau and Papua respectively. Reducing emissions in Riau would involve a cost of US$18.94/t$CO_2$eq for the first 154 million tCO_2eq (mainly from preventing the conversion of degraded forests to timber plantations). Reducing an additional 122 million tCO_2eq would cost US$56.34 per t$CO_2$eq, which arises from preventing the conversion of degraded forests to oil palm plantations (Figure 8.2). In Papua, reducing the first 2,998 million tCO_2eq would cost only US$0.56 per t$CO_2$eq (from preventing commercial logging in primary forests), whilst an additional reduction of 328 million tCO_2eq and 74 million tCO_2eq would cost US$12.90 per t$CO_2$eq (from preventing the conversion of degraded forests to timber plantations) and US$18.90 per t$CO_2$eq (from preventing the conversion of primary forests to oil palm plantations) respectively (Figure 8.3).

Using the derivation approach, the distribution of REDD+

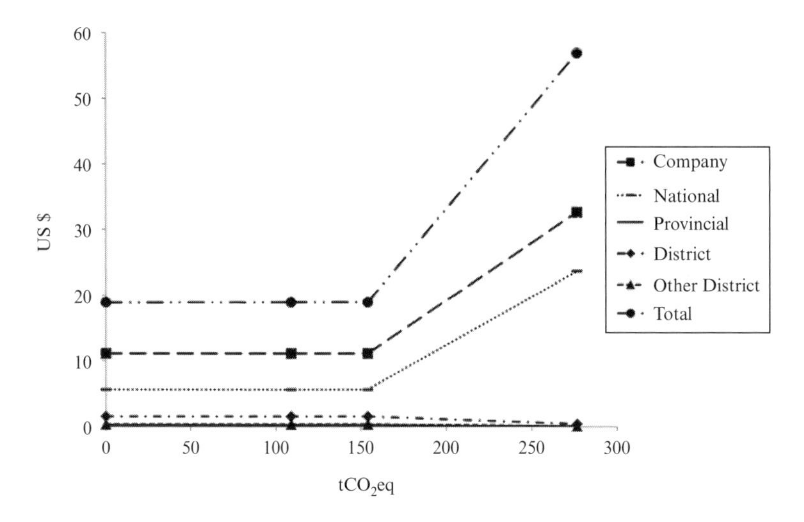

Figure 8.2 *The cost curve of reducing emissions from deforestation and forest degradation in Riau*

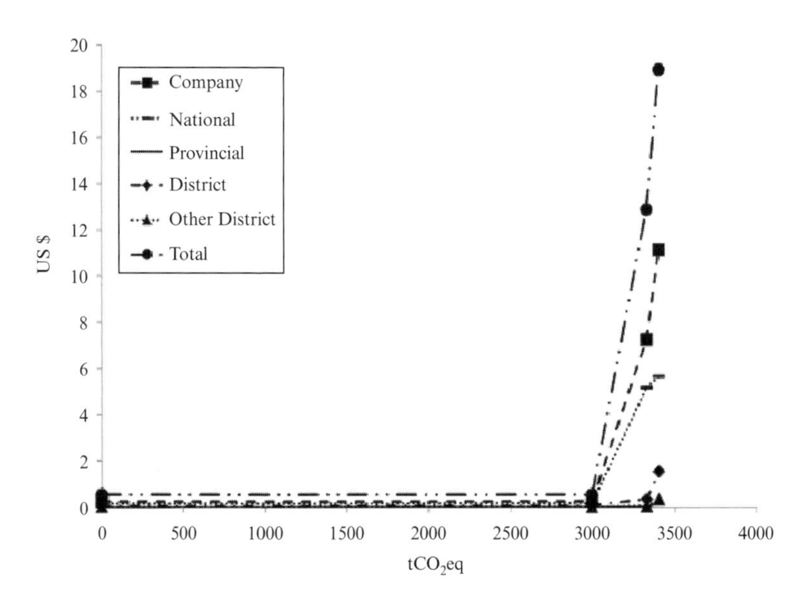

Figure 8.3 *The cost curve of reducing emissions from deforestation and forest degradation in Papua*

Table 8.5 Grant size for each government level using Approach 4 to determine RELs (US$ million)

Options	National	Provincial	Producing district	Other district
		Riau		
Cost reimbursement	784	30	217	50
Derivation Ministerial Decree 36/2009[1]	188	94	188	0
		Papua		
Cost reimbursement	305	12	150	46
Derivation Ministerial Decree 36/2009[2]	2,313	1,157	2,313	0

Notes:
1 Additional US$ 470 million for local communities.
2 Additional US$ 5.8 billion for local communities.

revenues is based on the assumed price of carbon and the shares of revenues allocated to local levels. Hence, the derivation approach in the distribution of benefits from REDD+ would set a flat rate per ton of carbon emissions reduced and ignore the opportunity costs of local governments from alternative land-use activities. If the carbon price were in the range of the 2011 carbon price in the European market (EUA), that was US$16–17,[4] the grant size for Papua would be significantly higher than the actual opportunity costs (Table 8.5). In contrast, the grant size for the district level in Riau that is determined using the derivation approach would be lower than that which is decided using the cost-reimbursement approach. Furthermore, the provincial government level in Riau could obtain a higher transfer when determining the size of IFTs using the derivation approach than the cost-reimbursement approach, because the share of revenues allocated for the provincial level is significantly higher under Ministerial Decree 36/2009 than it is if the actual opportunity costs are used (Table 8.5).

If all districts in Papua were required to reduce emissions by 50 per cent from the BAU level, most of the districts could reduce emissions at a cost of US$0.56 per tCO_2eq (from preventing commercial logging in primary forest). Some districts, such as Sarmi and Boven Digoel, would have higher opportunity costs to reduce 50 per cent of the emissions from the BAU level, as they would also need to prevent the conversion of primary forest to oil palm plantations (or agricultural activities), which would cost US$12.89 per tCO_2eq. In Riau, the cost of reducing 50 per cent of the emissions from the BAU level is approximately at US$18.96 per tCO_2eq (from preventing the conversion of degraded forests to timber plantations), except in a few districts such as Pekan Baru, Kuantan Singgingi and Dumai. In these districts, reducing 50 per cent of the emissions from the BAU level would also require preventing the conversion of degraded forest to oil palm plantations, which would cost US$56.34 per tCO_2eq.

Using the cost-reimbursement approach, districts with logged-over forests, such as in Riau, would receive higher revenues compared to their counterparts with intact primary forests, such as in Papua. Avoiding further conversion of logged-over areas is associated with higher opportunity costs than preventing the conversion of intact primary forests, as was discussed in Chapter 7. The conversion of logged-over forests is associated with higher opportunity costs because the alternative land-use activities in those areas are mainly timber plantations and oil palm plantations. These activities generate higher revenues than commercial logging that can merely take place in primary forest. Additionally, the total carbon stocks retained in logged-over forests are much lower than those of intact primary forests, which further increases the opportunity cost per unit carbon reduction (Palm et al., 1999).

Using the derivation approach, districts with more intact primary forests, which have not been exploited, would benefit more from REDD+ compared with their counterparts with more logged-over areas. Districts in Papua would receive approximately ten times the actual opportunity costs (Figure 8.4), whilst districts in Riau would receive transfers lower than their opportunity costs (Figure 8.5). With the assumed carbon price of US$16–17, Riau might not be interested in REDD+, particularly in forests currently under active concessions, although REDD+ may compete economically with other land-use activities in forest areas currently without active concessions as the government can retain all benefits that may accrue from REDD+. In forests currently without active concessions, the

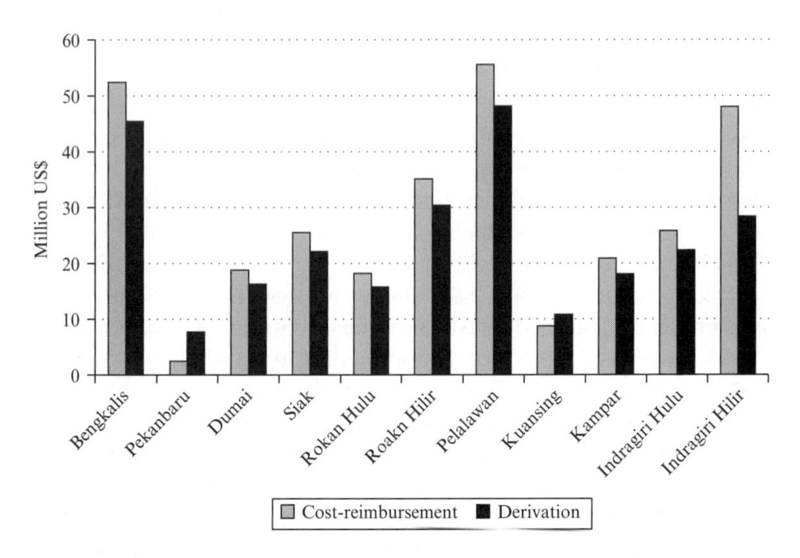

Figure 8.4 The size of IFTs allocated to district governments in Riau (US$)

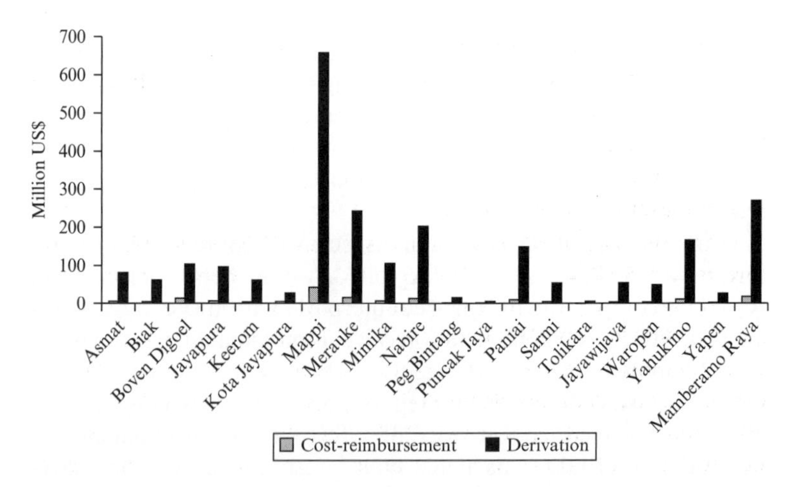

Figure 8.5 The size of IFTs allocated to district governments in Papua (US$)

portion of benefits for companies can be retained by the government entirely and distributed to different levels. For this reason, the government may want to conserve forest currently without active concessions and focus on maximizing revenues from productive activities in forest with active concessions. A total of 712,614 hectares and 3.8 million hectares of forests are currently without active concessions in Riau and Papua respectively.

Using both approaches, assuming that REDD+ would not provide compensation for the existing protected and conservation forests, districts such as Boven Digoel, Mappi, Merauke, Sarmi, Mamberamo and Asmat in Papua (Figure 8.4), would obtain higher transfers compared to their counterparts that do not have production forest areas such as Pegunungan Bintang and Puncak Jaya. Most districts with extensive conservation and protection forests in Papua are fiscally poor (i.e. Pegunungan Bintang and Puncak Jaya) because of reduced accessibility to forests and limited alternative land-use activities. In Riau, there is no clear relationship between fiscal capacity and the area of conservation and protected forest. The main income source in Riau is oil and the forestry and agricultural sectors do not play a significant role in the revenue composition (that is, fiscal capacity) of local governments. Districts that would obtain high revenues from REDD+ implementation in Riau are Bengkalis, Pelalawan and Indragiri Hilir (Figure 8.5). Bengkalis is one of the districts with very high fiscal capacity in Riau, while Pelalawan has low fiscal capacity.

8.5 DISCUSSION: THE DERIVATION *VS* THE COST-REIMBURSEMENT APPROACH

In this chapter, we demonstrate that it is possible to use both the derivation and the cost-reimbursement approaches to determine the IFT size to distribute REDD+ revenues. Using the cost-reimbursement approach, the IFT size allocated to district governments is just enough to cover the costs of reducing carbon emissions at the local level. The payment per ton of carbon for each district will vary depending on the alternative land-use activities and the carbon stocks retained in the forests. Local governments with low opportunity costs would receive a lower IFT size per ton of carbon compared with their counterparts with higher opportunity costs. The cost-reimbursement approach results in higher transfers to districts with more degraded forests than those with more primary forests.

Determining the IFT size using the cost-reimbursement approach would avoid an excessive producer surplus allocated to districts with low opportunity costs. The producer surplus is the area above the cost curve and below the price line (Hanley and Spash, 1993). At the international level, Cattaneo (2008), for instance, suggests that the producer surplus could be used to compensate for carbon stocks in standing forests to reduce leakage. Leakage, which occurs when the pressure of deforestation shifts from one location to others within a country, can result in no significant emission reductions from deforestation in the country overall. Providing incentives for all standing forests in a country could be expected to reduce leakage (ibid.). At the local level, a producer surplus can also be allocated to offset more emissions from deforestation and forest degradation in other localities with higher opportunity costs. However, localities with low opportunity costs would rather keep the producer surplus (as net profits), rather than allow the national government to keep it for offsetting more carbon (or the opportunity costs) in other locations. When deciding on the land-use allocation of a unit of land within their administrative boundaries, local governments would prefer to generate the highest return (profit) from the land, and would be less concerned about the total reductions of emissions achieved in the country.

This chapter focuses mainly on the opportunity costs, as they are thought to account for the largest share of the costs (Pagiola and Bosquet, 2008). However, the other costs, namely transaction and management costs, could be more than insignificant. The transaction costs reported in the literature are in the range of US$0.01–16.40 per tCO_2 (Wertz-Kanounnikoff, 2008), while the management costs, in the Brazilian Amazon for instance, were estimated at US$1–3 per hectare per year (Nepstad et al., 2009). Without the analysis of a specific location at the local level, it is problematic to determine whether the transaction costs of REDD+ depend mainly on the total CO_2 produced (therefore the quality of forests) and whether the management costs of REDD+ depend mainly on the total land areas as discussed in the literature. Should the management costs depend mainly on the total land area, then the results of this study would be unchanged: districts with more degraded forest would receive higher transfers than those with more primary forest. However, if transaction costs are influenced mainly by the total CO_2, then the costs of REDD+ in the primary forest (with higher CO_2 content) may be higher than in the degraded forest. However, the magnitude of the transaction costs may be different between regions in Indonesia depending on the political situation, capacity and leadership. Therefore, without a specific place-based analysis (which is beyond the scope of this research), it is not possible to establish how transaction and management costs might affect the total transfers for local governments to pursue REDD+. It is an issue requiring further research.

The derivation approach to determine the size of IFTs for local governments sets a flat rate per unit of avoided carbon emission, irrespective of the costs of REDD+ borne by local stakeholders. The size of IFTs is decided entirely by the defined percentage of revenues to be distributed to local levels and the price of carbon credits in the market. When the price of carbon is low, REDD+ can only attract the participation of low-cost districts, while a high carbon price would generate a producer surplus for low-cost producing districts. Hence, using the derivation approach, local governments should be allowed to decide voluntarily their participation in REDD+, based on their assessments of the costs and benefits of REDD+. Voluntary participation would allow the government to allocate lands for land-use activities that generate the highest return, either for productive activities or conservation (REDD+) activities, in order to maximise society's welfare as a whole.

Using the derivation approach, the grant size of IFTs allocated to the local levels is decided on the basis of a fixed percentage. In the case of Indonesia, Ministerial Decree 36/2009, for instance, stipulates that the district level obtains as much as 8 per cent of the total revenues generated from REDD+. In contrast, according to the results of the opportunity cost analysis reported in Irawan et al. (2013), the revenue distribution from oil palm plantations is currently centralised, with district governments obtaining a meagre portion of the total revenues captured by the government. Given the existing fiscal decentralisation in Indonesia, it is thus important to question whether distributing the revenues based on the existing incentive structure is sufficient to shift the interest of local governments in favour of conservation. Increasing the portion of revenues allocated for local governments would provide more incentives for local stakeholders to pursue REDD+. However, it would also compromise the total revenues retained by the national government. Therefore, if the national government is committed, then a larger portion needs to be assigned to local stakeholders at the expense of the national government's portion of revenues.

Using the derivation approach, the fluctuation of carbon prices will have no particular impact on the portion of transfers allocated to different levels vertically and between districts horizontally. The vertical distribution is influenced mainly by the percentage of revenue distribution, which is usually formalised by a government regulation; while the horizontal allocation of IFTs amongst districts is only influenced by the total carbon emissions avoided from deforestation and forest degradation in the respective districts. The changing carbon prices, however, will influence the willingness of local governments to participate in REDD+ (which is determined by the costs of REDD+).[5]

A carbon price of US$5 can offset the opportunity costs of reducing the first 2,998 million tCO_2eq emissions from preventing commercial logging in primary forests in Papua. To offset the additional 402 million tCO_2eq (from preventing the conversion of degraded forests to timber plantations) would cost US$ 12.9 per tCO_2eq. Therefore only when the price is around US$17 (as it was in the European Carbon Market in June 2011), could REDD+ offset all carbon emissions from REDD+ in Papua. In contrast, with a carbon price of US$17, only two districts in Riau, namely Pekanbaru and Kuantan Singingi, might be interested for REDD+. Only when the price of carbon is above US$20, can REDD+ offset the opportunity costs of reducing 154 million tCO_2eq emissions in Riau, mainly from preventing the conversion of degraded forests to timber plantations. Reducing an additional 122 million tCO2eq would cost US$56.34 per tCO_2eq, which cannot be offset under the likely price scenarios for carbon credits.

8.6 CONCLUSION

Building on the case of Indonesia with two contrasting provinces in terms of deforestation history and trends, this chapter has highlighted the importance of assigning sufficient financial resources to support conservation, and particularly REDD+, to local governments in decentralised countries. Intergovernmental fiscal transfers (IFTs) can be used as a means to distribute REDD+ revenues from the national level to local governments. In order to determine the IFT size to distribute REDD+ revenues, both the cost-reimbursement and the derivation approaches can be used. Using the cost-reimbursement approach, the IFTs distributed to local governments for pursuing REDD+ are determined entirely on the basis of the opportunity costs, which vary depending on land-use alternatives and the condition of the forests within a locality. There will, therefore, be an equity issue associated with this approach, as districts that have degraded their forest would receive higher revenues using this approach than those that have not pursued deforestation and forest degradation. Using the cost-reimbursement approach would also require an estimation of the costs of REDD+ for all localities, which may involve high transaction costs. In contrast, the distribution of REDD+ revenues amongst eligible district governments using the derivation approach ignores the opportunity costs of local governments from alternative land uses and focuses only on the market price of carbon credits and the share of revenues allocated to local levels. Localities with opportunity costs higher than the price of carbon credits should be allowed to refuse their participation in REDD+, while localities with low opportunity costs would be allowed to keep the producer surplus from reducing deforestation and forest degradation. Voluntary participation of local governments is therefore a prerequisite for this approach to succeed. Furthermore, using the derivation approach does not require an estimation of REDD+ costs for all districts, which will reduce the transaction costs of the implementation of REDD+.

Designing IFTs for channelling REDD+ revenues to local governments requires ecological indicators that differ from those that are usually used to determine the IFT size for biodiversity conservation discussed in the literature. Indicators that are commonly used to determine the size of IFTs for biodiversity conservation include the surface of protected areas, biodiversity indices and the total forest cover and geographical areas (e.g., Grieg-Gran, 2000; May et al., 2002; Köllner et al., 2002; Kumar and Managi, 2009; Ring, 2008b; Santos et al., 2012). REDD+ focuses on the additional forest conservation to avoid carbon emissions from deforestation and forest degradation and the enhancement of carbon stocks in forests. The distribution of REDD+ revenues should therefore consider indicators such as the historical deforestation rate, forest lands that can be legally converted to other land-use activities (the remaining carbon stocks in forests), and the alternative land-use activities.

Existing protected areas may not be eligible for compensation under REDD+ as compensation is only justifiable when a REDD+ measure can be shown to be contributing to an additional reduction of emissions and enhancement of carbon stocks. However, if there is no compensation for existing conservation forests, there will be an inequality implication for districts sustaining protected areas. With limited forest resources that could be legally exploited, those districts are often poorer than their counterparts with abundant production forests. If all the easily accessible forests are closed to exploitation, deforestation pressure could be shifted to protected and conservation forests. Financial allocations to support forest conservation would therefore be required. As a result, the distribution of REDD+ revenues should consider existing protected forests.

NOTES

1. This chapter is a slightly edited version of Irawan, S., Tacconi, L. and Ring, I. (2014). Designing Intergovernmental Fiscal Transfers for Conservation: The Case of REDD+ Revenue Distribution to Local Governments in Indonesia. *Land Use Policy*, 36, 47–59.
2. This rate is obtained based on the total area designated for timber plantations between 2000 and 2008, totalling 746,563 hectares (The Ministry of Forestry, 2008c).
3. 'Local community' probably refers to local stakeholder groups, although the decree does not provide a clear definition.
4. http://www.pointcarbon.com, accessed 5 June 2011.
5. The average price for offsets across the primary forest carbon markets rose from US\$3.8/tCO_2e in 2008 to US\$5.5/tCO_2e in 2010 (Diaz et al., 2011). Prices for EU allowances were trading at approximately €7/ton at the start of 2012, down more than 50 per cent from €15/ton at beginning of the 2011. CERs EU ETS were trading at just €3/ton at the beginning of January, down from €10/ton at the start of 2011. New Zealand carbon prices have similarly dropped to NZ\$7/ton at the start of 2012, compared to NZ\$20/ton at the beginning of last year. While stable for most of 2011, prices in North America's California carbon market at the end of 2011 have ended at US\$16/ton (TerraCarbon, 2012).

9. Conclusion

Local governments often obtain a portion of their revenues from land-use change and resource exploitation. Spatial externalities created by conservation result in local governments being reluctant to set aside productive lands for conservation, particularly when local people demand the conversion of ecosystems to pursue livelihood activities. Intergovernmental fiscal transfers (IFTs) can therefore play an important role in creating incentives for those governments to pursue conservation at the local level. This study has considered key elements of the design of IFTs for conservation, using the case of REDD+ revenue distribution in Indonesia. Studies on IFTs for conservation have focused mainly on the distribution formula (Köllner et al., 2002; Ring, 2008c; Kumar and Managi, 2009), while conditionality and accountability have received less attention. Moreover, research on IFTs for conservation, particularly in developing countries, has not considered the complex settings of government bureaucratic structures (Ring, 2008c; Kumar and Managi, 2009). Therefore, we hope that the findings of this study will contribute to the growing body of literature on IFTs for conservation. The study also provides a concrete example of how to design a mechanism to distribute revenues of REDD+ to subnational governments in developing countries.

In this final chapter, we reflect on the criteria for well-designed IFTs for conservation and the theoretical and practical implications for the design of IFTs for REDD+ revenue distribution that arise from the analysis presented throughout the book. The chapter first considers the contribution of the findings of this study to the literature on IFTs for conservation and the policy analysis framework for developing environmental policy. The discussion then proceeds with the design of IFTs for REDD+ revenue distribution and situates it in the different phases of the implementation of REDD+ in Indonesia. As discussed in Chapter 3, REDD+ may be implemented in three phases: strategy development, readiness and full implementation phases. Different funding mechanisms may be involved in each phase (Vatn and Angelsen, 2009; Meridian Institute, 2009), hence it is important to explore when to use IFTs for REDD+ revenue distribution within the different implementation phases. We conclude with a discussion of some limitations of the study and provide recommendations for future research.

9.1 CRITERIA FOR THE DESIGN OF IFTs FOR CONSERVATION

Designing IFTs for forest conservation requires several considerations concerning: (i) the characteristics of the distribution formula, conditionality and accountability mechanisms; (ii) the technical capacity of local governments that will implement the IFTs; and (iii) the local forest governance conditions that influence local governments' interest in, and commitment to, forest conservation. Through a summary of the analysis contained in the previous chapters, here we provide some reflections on the technical considerations for the design of IFTs, while the technical capacity of local governments and local forest governance are discussed in the next section.

The first consideration for designing IFTs for conservation is to clearly define the objectives to be achieved. Second, the distribution formula should be neutral, understandable and not able to be influenced by local governments through the manipulation of their expenditure or tax decisions. Third, IFT mechanisms should ensure the autonomy and independence of local governments to set the priorities to respond to local needs. Fourth, the design should ensure accountability and transparency in the process of allocating transfers from the national to local level as well as in the implementation at the local level. Finally, the design should be simple, easy to administer and involve the lowest possible administration and transaction costs.

In relation to the objectives of IFTs, the literature suggests that they should improve the effectiveness and efficiency of the provision of public services at the local level (Bird, 1999; Bird and Smart, 2002; Shah, 2006). This study suggests two main objectives for IFTs for REDD+ which can be applied to IFTs for conservation in general: (i) ensuring that conservation will not compromise local governments' fiscal capacity to provide basic public services; and (ii) financing conservation or forest-related services at the local level. These objectives correspond to the cost elements of conservation incurred by local governments: opportunity, management and transaction costs (Naidoo and Ricketts, 2006; Naidoo et al., 2006). The first objective of IFTs for conservation is in line with the compensation for the opportunity costs of conservation. If the opportunity costs are not compensated properly, the revenue stream of local governments is affected, which impacts on their capacity to provide basic public services. The second objective relates to the management and transaction costs of conservation. The IFTs should be allocated to ensure local governments have resources to finance specific conservation activities.

For the first purpose of IFTs for conservation, two approaches can be used to determine the size of IFTs to local governments, namely the cost-reimbursement and the derivation approaches. The cost-reimbursement approach has been used in Brazil and Portugal and has been proposed in a number of other countries to correct the spillover effect of conservation (e.g., Ring, 2008b; Köllner et al., 2002; Ring, 2008c; Santos et al., 2012; Grieg-Gran, 2000; May et al., 2002). The cost reimbursement approach can be used to correct spillover effects of the provision of public services by providing a unit subsidy to local governments as much as the value at the margin of the spillover benefits they create (Oates, 1972, p. 66). Although these studies do not estimate the actual spillover costs or benefits per locality, they use ecological indicators as a proxy for determining the costs of conservation, such as the extent of protected areas, the ecological functions of forests, species diversity, and the total land area (Grieg-Gran, 2000; May et al., 2002; Köllner et al., 2002; Ring, 2008b; Kumar and Managi, 2009). In the case of REDD+, the determination of opportunity costs can use some indicators, including alternative land-use activities, the historical rates and practices of deforestation and the unit of environmental services that can be generated (such as carbon emissions reduced and carbon stocks enhanced). These indicators are also relevant for determining the size of IFTs for additional forest areas designated as conservation areas.

Since environmental services from conservation might result in the generation of revenues, such as is potentially the case for REDD+ and other types of payments for environmental services (PES), environmental services can be regarded as being similar to forest commodities such as timber and other non-timber forest products. This new perspective requires a different approach to determine the size of IFTs to support conservation. As with other forest products, the size of IFTs for local governments to support conservation could be decided using the derivation approach. This approach allocates a portion of benefits to local governments based on a specified percentage of taxes or fees collected from conservation activities and their corresponding environmental services, such as carbon emissions reduced and carbon stocks enhanced. Using this approach, the size of IFTs allocated to local governments ignores the costs of conservation incurred at the local level. In the case of REDD+, for instance, local governments could be expected to be interested in participating when the costs of conservation are lower than the revenues from REDD+. Local governments should then be allowed to decide whether to participate in a REDD+ scheme on the basis of their assessment of the costs and benefits of doing that.

For the second purpose, the cost-reimbursement (or formula) approach can be used to determine the size of IFTs to finance public services related to conservation and forest management at the local level. Using quantitative criteria, the distribution formula can be developed based on the estimation of the cost to provide forest-related services to local government units. Management and transaction costs can be estimated based on activities that should be performed by local governments to support the implementation of REDD+ at the local level. They include hiring staff, building necessary infrastructure and community development for people living in or near surrounding forests. The management costs of forest conservation may also need to take into account regional disparities, such as fiscal capacity.

Both the cost-reimbursement and the derivation approaches are neutral, as local governments cannot influence the size of IFTs by manipulating their expenditure or tax decisions. The cost-reimbursement approach can be used when local governments are obliged to provide a certain level of conservation at the local level as part of a determined minimum service standard. Using this approach, the national government can reimburse the costs of conservation borne by local governments. On the other hand, when local governments are required to provide additional conservation areas beyond the existing ones (regulated by law), then the derivation approach could be used.

The findings of this study also enhance our understanding of conditionality of IFTs in conservation, a rather contentious issue that has not received sufficient attention in the existing literature (e.g. Ring, 2008b; Ring, 2008c; Köllner et al., 2002; Santos et al., 2012; Grieg-Gran, 2000; May et al., 2002). While the flexibility to spend funds at the local level is important to allow local governments to pursue local priorities, a degree of conditionality is also required to ensure that spending takes place on the services in question (Ring, 2008c). As with the distribution formula, the conditionality of IFTs should also consider the opportunity, management and transaction costs of conservation. Funding aimed at compensating opportunity costs should be transferred with flexibility for local governments to decide the activities and policies to be pursued. This should also allow local governments to use IFTs for general budget support and to finance basic public services in other sectors. However, funding for the management costs of conservation should be transferred with conditions aimed at ensuring that the funds are used to finance conservation activities. Interviews with government officials at all levels in Indonesia revealed that some earmarking of IFTs was preferred in conservation for political and administrative reasons. If they are not specifically earmarked, those funds could be diverted to other development priorities because local governments face pressures to address the immediate needs of local people.

Accountability of IFTs should consider the purpose and conditionality of the transfers. After receiving REDD+ payments, local governments must be accountable to those who pay for them and those who benefit from them (Shah, 2006; Bird, 1999). Bird (1999) argues that the provision of sufficient information to local constituents is important to ensure transparency in the process. The accountability aspect discussed in this study focuses on the obligation of local governments as the recipients of IFTs to perform the responsibilities conferred. Output-based accountability can be applied to the component of IFTs aimed at compensating for the opportunity costs, where the transfer can only be made when the recipients have achieved the prescribed targets and performance. In addition, transfers to finance specific services to be performed by local governments (related to transaction and management costs of conservation) can use a conditional grant with an input-based accountability mechanism. This accountability measure can include the monitoring of local governments' spending on specific forest-related activities and services. Disbursement of the funds can be conducted at the beginning of conservation activities and monitored at the end of the fiscal year or other relevant period. To ensure that technical performance is achieved, the agencies responsible for conservation could monitor conservation activities implemented at the local level.

Finally, in order to ensure that IFTs are implemented with the lowest possible administration and transaction costs, it is important to consider the existing capacity of local governments to manage public resources at the local level. This is addressed below, together with factors relating to the effectiveness of IFTs.

9.2 CAPACITY AND EFFECTIVENESS CONSIDERATIONS

In designing IFTs for conservation, the political and administrative factors that influence their implementation within a complex bureaucratic environment, particularly the technical capacity of implementing agencies, need to be considered (Williams, 1975, p. 558). If international payments for conservation were to flow from the international to the local level, a sharp increase in financial resources could occur. This could be problematic given the limited capacity of local governments in managing financial resources. For instance, local governments in Indonesia have accumulated substantial unspent balances due to their low capacity in public service delivery (Alisjahbana, 2005; Lewis and Oosterman, 2009).

The technical capacity to implement IFTs, including the capacity to collect taxes and fees and/or to spend the funds transferred, differs across local governments. In terms of revenue sharing, the size of IFTs allocated to the local level depends on taxes and fees collected in relation to conservation and environmental services generated at the local level. Although the national government can collect taxes and fees without the involvement of local governments, local governments' participation is important to ensure ownership and transparency in the revenue distribution. In the case of REDD+ implementation, for instance, transfers could be made based on the reduction of emissions and carbon stock enhancement achieved by a locality. The involvement of local governments in the collection of data related to emission reduction and carbon stock enhancement may be necessary to ensure transparency in the allocation of IFTs to the local level. In the case of payments for environmental services, local governments need to be aware of the services provided by their localities and the amount of compensation they are entitled to. Hence, the use of revenue sharing in conservation would require new skills for governments at all levels, not only in the management of funds but also in data collection and monitoring related to the environmental services produced. Alternatively, an unconditional grant could be implemented as an option to the revenue-sharing mechanism. The unconditional grant does not require the participation of local governments in the collection of taxes and fees. The size of the transfer is decided by the national government, based on quantitative criteria monitored directly by the national government.

As with an unconditional grant, a conditional grant does not require the involvement of local governments in collecting taxes or fees. A conditional grant to finance management activities related to conservation at the local level can be implemented without new skills being required, except the routine management activities of conservation. Due to low capacity in financial management, clear guidance from the national government is preferable to prevent mistakes in spending of the funds transferred, which could lead to corruption charges. However, guidelines developed by a national government about how to spend funds at the local level need to take into account local conditions, and leave some flexibility for local officials to pursue activities that are important for their localities. The conditions of forests in each locality vary widely, as do the economic and social situations. If they do not allow for variations between localities, guidelines provided by the national governments are often not implementable, as discussed in Chapter 6.

In relation to effectiveness, it is important to consider the wider political economy of land-use change, as ultimately IFTs aim to change the behaviour of local public actors in order to achieve improved conservation outcomes. Legal land-use change and forest exploitation generate income for national and local governments. These activities also provide employment and other economic benefits that are important for local people's livelihoods. People in different localities might have different beliefs and values towards deforestation and forest degradation. The beliefs and values of local government officials seem to be heavily influenced by the local people in Indonesia, as demonstrated by the fact that local officials considered themselves to be the first line of bureaucrats who cater for local people's needs. Moreover, within a decentralised country, public officials at different government levels may not have common interests about forest resources and conservation. Development policies decided at the national level might also not reflect the priorities of local decision makers. This implies that there is a range of values about forest resources, with possible differences between local governments and local communities, as well as between local governments and their national counterparts.

Due to the range of values and beliefs around land-use change, several factors need to exist together with IFTs to ensure their successful implementation. First, local governments would need to have the authority to decide voluntarily whether to allocate additional forests for conservation. Providing greater authority to local levels would not always lead to forest conservation (Tacconi, 2007). However, as discussed in Chapter 5, local decision makers should be allowed to decide on behalf of the local constituents whether to allocate productive lands within their administrative boundaries for conservation. Since additional conservation may not necessarily lead to additional environmental services, particularly for local residents (Pattayanak et al., 2010), government officials in developing countries need to assess the costs and benefits of allocating more land for conservation, as discussed in Chapter 7. Second, IFTs for conservation should be accompanied by mechanisms to create incentives for local communities to favour conservation, since the beliefs and values of local government officials concerning land-use allocation are influenced by the local people's demands. For instance, a specific portion of benefits could be paid to local communities through payment for environmental services schemes. Third, upward and downward accountability measures are required to prevent the misuse of funds. NGOs' protests and coercive action by national governments can also help prevent the misuse of funds transferred to the local level. In the case of Indonesia, for instance, prosecution of corrupt activities has caused local officials to be more cautious in managing public resources.

9.3 THE PROPOSED DESIGN OF IFTs FOR REDD+ REVENUE DISTRIBUTION IN INDONESIA

This study has discussed the perspectives of government officials who would be involved in the development and implementation of

IFTs in Indonesia about the options for IFTs for REDD+ revenue distribution. As noted earlier, the first consideration when developing options for IFTs for REDD+ revenue distribution is the purpose of the transfers. IFTs for REDD+ revenue distribution need to serve two main purposes: (i) ensuring that REDD+ measures will not affect local governments' fiscal capacity to deliver public services; and (ii) financing forest-related services at the local level to ensure the successful implementation of REDD+. According to the perspectives of government officials, the two preferred mechanisms to distribute REDD+ revenues were the revenue-sharing mechanism and conditional grants. Each of the mechanisms can serve different purposes of IFTs for REDD+ revenue distribution.

A revenue-sharing mechanism can be used to ensure that REDD+ measures will not affect local governments' fiscal capacity to deliver public services. An alternative to the revenue-sharing mechanism is an unconditional grant. Government officials interviewed in this study did not specifically mention unconditional grants as their preferred mechanism because the unconditional grant mechanism currently implemented in Indonesia (known as the General Allocation Fund) is not considered suitable for REDD+ revenue distribution. Using the existing unconditional grant mechanism to transfer REDD+ revenue would distort the purpose of the grant, which mostly aims to finance the salaries of civil servants at the local level, and may not create a strong and direct link between the funds transferred with activities for reducing deforestation implemented at the local level (Chapter 6). The unconditional grant proposed here is a specific lump sum transfer created for REDD+ revenue distribution, which is separate from the existing unconditional grant already being implemented.

The revenue-sharing mechanism and the unconditional grant do not have any conditionality requirements. They are distributed to local stakeholders without any specific conditions, although they differ in terms of the distribution formula, as discussed below. Both mechanisms can be used to compensate the forgone taxes and fees, which are important to finance routine expenditures and public services in other sectors, as discussed in Chapters 7 and 8. Local governments can then decide the use of the funds based on activities that are considered important for their localities. If these mechanisms were to be used to distribute REDD+ revenues, the transfer would have to carry output conditionality (or performance-based accountability), where the transfers can only be made when recipients have achieved the prescribed targets (Shah, 2006). In the case of REDD+, a transfer could be disbursed once emission reductions have been measured, reported and verified (Table 9.1).

The revenue-sharing mechanism usually applies the derivation approach, while the size of an unconditional grant can be determined using the formula approach (see Chapter 8). In the case of REDD+, determining the size of IFTs for local governments using the derivation approach means setting a flat rate of compensation per unit of emission reduction. Thus the size of IFTs depends on the total emission reductions achieved in a locality and the percentage of revenues allocated to the local level. Using the formula approach, the size of the transfer to local governments is determined on the basis of the costs faced by local governments to implement REDD+. Local governments with low opportunity costs would receive lower compensation for each unit of emissions reduced than would their counterparts with higher opportunity costs.

The financing of specific forest-related services at the local level to ensure successful REDD+ implementation could use a conditional grant. The use of the funds to finance specific activities for the implementation of REDD+ at the local level needs to be earmarked. The national government could provide prescriptions on activities that need to be financed at the local level. However, local governments should also be provided with the flexibility to pursue activities that are most required in their localities. Conditional grants can specify the type of expenditure that can be financed (input-based or expenditure conditionality) and the disbursement usually requires local governments to report to the national government on a regular basis. Conditional grants would therefore allow for upfront financing to invest in REDD+ activities during the initial period of implementation (Table 9.1).

This study, however, did not assess the size of IFTs required to finance specific services at the local level related to REDD+ implementation, due to the lack of data on REDD+ management and transaction costs. The national government could use a cost reimbursement (formula) approach to determine the grant size to finance specific services at the local level (Bird, 1999; Bahl, 2000; Bird and Smart, 2002; Shah, 2006). The formula approach can apply quantitative criteria to estimate the amount needed to provide REDD+ related services to eligible local government units. In the

Table 9.1 Options for REDD+ revenue distribution

	Revenue sharing	Unconditional grant	Conditional grant
Purpose of IFTs	Ensuring that REDD+ measures will not affect local governments' fiscal capacity to deliver public services	Ensuring that REDD+ measures will not affect local governments' fiscal capacity to deliver public services	Financing forest-related services at the local level to ensure the successful implementation of REDD+
Grant size	Percentage of national taxes or fees	Formula approach based on the opportunity costs	Distribution formula based on costs of providing services at local levels
Distribution formula	Percentage of revenues allocated to district level, total emission reduction, and the price of REDD+ carbon credit	Opportunity costs per ton of carbon and total emission reduction	Cost elements of providing related services (including physical indicators, regional characteristics, population)
Conditionality	Unconditional	Unconditional	Earmarked for a specific sector with a tentative list of activities
Accountability	Performance accountability (based on emission reductions achieved)	Performance accountability (based on emission reductions achieved)	Input accountability (spending – quarterly report)

case of REDD+, estimation of the transaction and management costs would require a clear definition of roles and responsibilities to be performed by local governments in its implementation.

As discussed in Chapter 6, officials from forestry, finance and planning agencies at the national, provincial and district levels had different views on the options for the design of an IFT mechanism. Differences in perspectives are more significant between agencies than they are between government levels. Officials from forest agencies are mostly concerned with generating revenues from forest resources and implementing forest-related services. Hence, they favoured a revenue-sharing mechanism for REDD+ revenue distribution. They also preferred flexibility in terms of the use of the funds. Chapter 6 noted that the guidelines about the use of the reforestation fund do not consider local conditions. As a result, many forest rehabilitation activities are not implementable. In contrast, the finance agencies' officials thought that a conditional transfer is preferable to revenue sharing. Those officials are responsible for the timely disbursement of funds. They were, therefore, more concerned with the administration of the funds and less with the technical performance of the service delivery. Conditional grants are considered easier to administer. Clear guidelines about how funds transferred to the local level should be spent can prevent any mistakes in using the funds, particularly when capacity in financial management is generally low at the local level (Chapter 6). Officials from the planning agencies at all levels appeared to be mostly concerned about overall economic development, particularly the welfare of local communities. Therefore, they favoured a revenue-sharing mechanism to ensure that REDD+ benefits could also be spent for basic service provision. The main difference in views between officials at different government levels is about the distribution formula and the grant size of REDD+ revenues. As might be expected, local government officials believe that the share of REDD+ revenues to be distributed to the local governments should be higher than the existing share from taxes on resource extraction.

In discussing the preferred mechanisms of IFTs for REDD+, government officials considered the existing capacity of local governments to manage public revenues at the local level. As previously mentioned, conditional grants are considered easy to administer as they are transferred with clear prescriptions on how they can be spent. In contrast, it was noted that using revenue sharing to distribute REDD+ revenues requires a certain capacity of local governments to be involved in the data collection and monitoring of emission reductions or carbon stock enhancement. The use of revenue-sharing mechanisms for REDD+ revenue distribution may well be constrained by the existing capacity of local government.

9.4 SOME LIMITATIONS OF THIS STUDY

This study aimed to cover all the elements of well-designed IFTs for conservation, which in turn reduced somewhat the depth of analysis of the individual elements. For instance, using a larger sample may help understand the impacts of the distribution formula on localities whose characteristics differ from those of the two sample provinces selected for this study. Another issue is that a single-country case study approach was used. Caution must therefore be applied, as generalisation of the findings could be problematic. However, restricting this study to one country enabled us to demonstrate in detail how an IFT mechanism may be studied and designed, thus facilitating replication in other decentralised countries that may be seeking to implement nationwide conservation initiatives.

In relation to the opportunity costs analysis, the study applied a local empirical approach, which has the advantage of being transparent, yielding relatively accurate estimates and being simple to interpret. However, extrapolation should be done cautiously, as the results are specific to a particular region (Richards and Stokes, 2004; Wertz-Kanounnikoff, 2008). Moreover, as discussed in Chapter 7, the financial assumptions used in the estimation of the opportunity costs can result in wide variations. Any extrapolation, therefore, needs to pay attention to these assumptions. Finally, the analysis relied, to a certain extent, on secondary data, particularly on carbon stocks. An extensive literature review was conducted to obtain the most reliable data and a sensitivity analysis was also performed. However, future analyses, both in Indonesia and in other countries, should seek to use data from ground measurements of forest carbon stocks to improve the accuracy of the cost estimates. Spatial information would also improve the estimates of the opportunity costs by contributing to a more accurate assessment of the land-use activities for a specific location and the total carbon emitted due to such activities.

Finally, this study has focused on estimating the opportunity costs of forest conservation for REDD+ without considering management and transaction costs. Addressing these costs would improve the accuracy of the findings in relation to the overall costs of REDD+ and, among other things, improve the estimates of the funds required by local governments to implement REDD+.

9.5 FUTURE RESEARCH

The previous section has identified some possible limitations of this study, and specific issues that may be considered by future studies concerned with the design of IFTs for conservation. Let us now consider some broader challenges for future research.

There is a need to consider performance-based IFTs where the transfers would be made on the basis of local governments' performance in implementing conservation policies. In the public finance literature, several studies analyse performance-oriented IFTs in public service delivery, where result-based accountability is applied to measure the success of the implementation of IFTs. Result-based accountability is advocated in order to move from monitoring local governments' spending to monitoring the results achieved in the efficient delivery of public services (Shah, 2006). However, there is currently no research on performance-based IFTs for conservation. This type of IFT will have implications for the transfer schedule of IFTs to eligible local units, as performance can only be assessed after a certain period of time. The size of the transfers would change when the discounting factor (or the interest rate) is considered, particularly when a payment is disbursed after localities achieve certain performance in conservation policies.

In relation to the distribution formula for IFT revenues, this study has focused only on the empirical analysis of the opportunity costs due to the paucity of data on the potential transaction and management costs of REDD+ in Indonesia. Some studies have researched the management and transaction costs of conservation (Naidoo and Ricketts, 2006; Naidoo et al., 2006). In the case of REDD+, specific management activities should be implemented, such as collecting and monitoring data on carbon emission reductions and carbon stock enhancement. Although some studies have examined the management and transaction costs of REDD+ (Nepstad et al., 2009; Pagiola and Bosquet, 2008), the estimation of REDD+ transaction and management costs specifically incurred by local governments will require a clear definition of local government's roles and responsibilities in the implementation of REDD+. However, this aspect has not received the necessary attention yet. The transaction costs of REDD+ reported in the literature are in the range of US$0.01–16.40 per tCO_2 (Wertz-Kanounnikoff, 2008), while the management costs, in the Brazilian Amazon for instance, were estimated at US$1–3 per hectare per year (Nepstad et al., 2009). Without the analysis of specific locations, it is problematic to determine whether the transaction and management costs of REDD+ depend on the total CO_2 produced or the total land areas included in a REDD+ scheme. The magnitude of the transaction costs may differ between regions, depending on the political situation, capacity and leadership. Therefore, future studies should consider analysing how transaction and management costs might affect the total transfers for local governments to pursue REDD+. Such research would enhance knowledge about the economic feasibility of REDD+ for localities in developing countries. Consideration of transaction and management costs is also relevant for the design of IFTs focused on the conservation of ecosystems other than forests.

References

Agrawal, A., Chhatre, A. and Hardin, R. (2008) Changing governance of the world's forests. *Science*, 320, 1460–1462.

Alisjahbana, A. (2005) Does Indonesia have the balance right in natural resource revenue-sharing? In Resosudarmo, B. (ed.) *The Politics and Economics of Indonesia's Natural Resources*. Singapore, Institute of Southeast Asian Studies.

Andersson, K. (2003) What motivates municipal governments? Uncovering the institutional incentives for municipal governance of forest resources in Bolivia. *The Journal of Environment and Development*, 12, 5–27.

Andersson, K. (2004) Who talks with whom? The role of repeated interactions in decentralized forest governance. *World Development*, 32, 233–249.

Andersson, K. and Gibson, C. C. (2007) Decentralized governance and environmental change: Local institutional moderation of deforestation in Bolivia. *Journal of Policy Analysis and Management*, 26, 99–123.

Andersson, K., Gibson, C. C. and Lehoucq, F. (2004) The politics of decentralized natural resource governance. *Political Science and Politics*, 37, 421–426.

Andersson, K., Gibson, C. C. and Lehoucq, F. (2006) Municipal politics and forest governance: Comparative analysis of decentralization in Bolivia and Guatemala. *World Development*, 34, 576–595.

Andersson, K., Evans, T., Gibson, C. C. and Wright, G. (2010) Decentralization and deforestation: Comparing local forest governance regimes in Latin America. In Andreas Duit (ed.) *Mapping the Politics of Ecology: Comparative Perspectives on Environmental Politics and Policy*. Cambridge, MA: MIT Press.

Andrianto, A., Obidzinski, K., Wajdi, F. and Tetuka, B. (2008) Deforestation and forest degradation in Papua post-new order regime. Bogor, Indonesia: Center for International Forestry Research.

Angelsen, A., Streck, C., Peskett, L., Brown, J. and Luttrell, C. (2008) What is the right scale for REDD? The implications of national, subnational and nested approaches. *Info Brief*. Bogor, Indonesia: Center for International Forestry Research.

Ashton, R., Boer, R., Cosler, P., DeFries, R., El-Ashry, M. et al. (2008) *How to Include Terrestrial Carbon in Developing Nations in the Overall Climate Change Solution*. Boston, MA: The Terrestrial Carbon Group.

Badan Pusat Statistik (BPS) Papua (2009) Papua Dalam Angka 2008. Jayapura.

Bahl, R. (1999) Implementation rules for fiscal decentralization. *International Studies Program Working Paper 99-1*. Atlanta, Andrew Young School of Policy Studies, Georgia State University.

Bahl, R. (2000) Intergovernmental transfers in developing and transition countries: Principles and practice. *Urban and Local Government Background Series No. 2*. Washington, DC: The World Bank.

Bahl, R. and Linn, J. (1992) *Urban Public Finance in Developing Countries*. New York: Oxford University Press.

Bahl, R. and Wallace, S. (2005) Public financing in developing and transition countries. *Public Budgeting and Finance*, 25, 83–98.

Bahl, R. and Wallace, S. (2007) Intergovernmental transfers: the vertical sharing dimension. In Martinez-Vazquez, J. and Searle, B. (eds) *Fiscal Equalization*. New York: Springer.

Barr, C. (2000) Profits on paper: The political-economy of fiber, finance, and debt in pulp and paper industries. Bogor, Indonesia: Center for International Forestry Research.

Barr, C., Resosudarmo, I. A. P., Dermawan, A., Mccarthy, J. F., Moeliono, M. and Setiono, B. (2006) Decentralization of forest administration in Indonesia: Implications for forest sustainability, economic development and community livelihoods. Bogor, Indonesia: Center for International Forestry Research.

Barr, C., Dermawan, A., Purnomo, H. and Komarudin, H. (2009) Financial governance and lessons from Indonesia's reforestation fund. Bogor, Indonesia: Center for International Forestry Research.

Bennett, A. and Elman, C. (2006) Qualitative research: Recent developments in case study methods. *Annual Review of Political Science*, 9, 455–476.

Bird, N. (2005) Considerations for choosing an emission target for compensated reductions. In Moutinho, P. and Schwartzman, S. (eds) *Tropical Deforestation and Climate Change*. Washington DC: Instituto de Pesquisa Ambiental da Amazônia.

Bird, R. (1999) *Transfer and Incentives in Intergovernmental Fiscal Relations*. Washington DC, The World Bank.

Bird, R. (2001) *Intergovernmental Fiscal Relations in Latin America: Policy Design and Policy Outcomes*. Washington DC, Inter-American Development Bank.

Bird, R. and Ebel, R. (2007) *Fiscal Fragmentation in Decentralized Countries: Subsidiarity, Solidarity, and Asymmetry*. Cheltenham, Edward Elgar Publishing.

Bird, R. and Smart, M. (2002) Intergovernmental fiscal transfers: International lessons for developing countries. *World Development*, 30, 899–912.

Blatter, J. and Blume, T. (2008) In search of co-variance, causal mechanisms or congruence? Towards a plural understanding of case studies. *Swiss Political Science Review*, 14, 315–356.

Boucher, D. (2008) *Out of the Woods: A Realistic Role for Tropical Forests in Curbing Global Warming*. Washington, DC: Union of Concerned Scientists.

Box, R. (1999) Running government like a business. *The American Review of Public Administration*, 29, 19–43.

Bradbury, M. D. and Waechter, G. D. (2009) Extreme outsourcing in local government. *Review of Public Personnel Administration*, 29, 230–248.

Bräutigam, D. (2002) Building leviathan: Revenue, state capacity, and governance. *IDS Bulletin*, 33, 1–17.

Busch, J., Strassburg, B., Cattaneo, A., Lubowski, R., Bruner, A. et al. (2009) Comparing climate and cost impacts of reference levels for reducing emissions from deforestation. *Environmental Research Letter*, 4, 1–11.

Busch, J., Lubowski, R., Godoy, F., Yusuf, A., Austin, K. et al. (2011) Structuring national and sub-national economic incentives to reduce emissions from deforestation in Indonesia. *Working Papers in Economics and Development Studies*. Bandung, Indonesia: Department of Economics, Padjadjaran University.

Butler, R. A., Koh, L. P. and Ghazoul, J. (2009) REDD in the red: Palm oil could undermine carbon payment schemes. *Conservation Letters*, 2, 67–73.

Cacho, O. J., Marshall, G. R. and Milne, M. (2005) Transaction and abatement costs of carbon-sink projects in developing countries. *Environment and Development Economics*, 10, 597–614.

Cattaneo, A. (2008) How to distribute REDD funds across countries? A stock-flow mechanism. Falmouth, MA: Woods Hole Research Center.

Cattaneo, A. (2011) Robust design of multiscale programs to reduce deforestation. *Environment and Development Economics*, 16, 455–478.

Cheema, G. S. and Rondinelli, D. A. (1983) *Decentralization and Development Policy Implementation in Developing Countries*, Beverly Hills: Sage.

Coffey, A. and Atkinson, P. (1996) *Making Sense of Qualitative Data*, London: Sage.

Colchester, M., Jiwan, N., Andiko, Sirait, M., Firdaus, A. et al. (2006) *Promised Land: Palm Oil and Land Acquisition in Indonesia – Implications for Local Communities and Indigenous Peoples*. Bogor, Indonesia: Forest Peoples Programme, Perkumpulan Sawit Watch, HuMA and the World Agroforestry Center.

Contreras-Hermosilla, A. and Fay, C. (2005) *Strengthening Forest Management in Indonesia Through Land Tenure Reform: Issues and Framework Action*. Washington, DC: Forest Trends and Bogor, Indonesia: World Agroforestry Center.

Conyers, D. (1983) Decentralization: The latest fashion in development administration? *Public Administration and Development*, 3, 97–109.

Corbera, E., Estrada, M. and Brown, K. (2010) Reducing greenhouse gas emissions from deforestation and forest degradation in developing countries: Revisiting the assumptions. *Climatic Change*, 100, 355–388.

Dalle, S. P., De Blois, S., Caballero, J. and Johns, T. (2006) Integrating analyses of local land-use regulations, cultural perceptions and land-use/land cover data for assessing the success of community-based conservation. *Forest Ecology and Management*, 222, 370–383.

Davoodi, H. and Zou, H. F. (1998) Fiscal decentralization and economic growth: A cross-country study. *Journal of Urban Economics*, 43, 244–257.

De Mello, L. R. (2000) Fiscal decentralization and intergovernmental fiscal relations: A cross-country analysis. *World Development*, 28, 365–380.

Diaz, D., Hamilton, K. and Johnson, E. (2011) *State of the Forest Carbon Markets 2011: From Canopy to Currency*. Washington, DC: Forest Trends.

Ebeling, J. and Yasué, M. (2008) Generating carbon finance through avoided deforestation and its potential to create climatic, conservation and human development benefits. *Philosophical Transactions of the Royal Society B: Biological Sciences*, 363, 1917–1924.

Eliasch Review (2008) *Climate Change: Financing Global Forests*. London: HMSO.

Fadliya and Mcleod, R. H. (2010) Fiscal transfers to regional governments in Indonesia. *Departmental Working Papers*, Arndt-Corden Department of Economics, The Australian National University, Canberra.

Falleti, T. G. (2005) A sequential theory of decentralization: Latin American cases in comparative perspective. *American Political Science Review*, 99, 327–346.

Falleti, T. G. and Lynch, J. F. (2009) Context and causal mechanisms in political analysis. *Comparative Political Studies*, 42, 1143–1166.

Feldman, M. and Martin, R. (2005) Constructing jurisdictional advantage. *Research Policy*, 34, 1235–1249.

Fisher, B., Edwards, D. P., Giam, X. and Wilcove, D. S. (2011) The high costs of conserving Southeast Asia's lowland rainforests. *Frontiers in Ecology and the Environment*, 9, 329–334.

Fischer, F. (2003) *Reframing Public Policy: Discursive Politics and Deliberative Practices*. Oxford: Oxford University Press.

Føllesdal, A. (1998) Survey article: Subsidiarity. *Journal of Political Philosophy*, 6, 190–218.

Forest Watch Indonesia and Global Forest Watch (2002) *The State of the Forest Indonesia*. Jakarta, World Resources Indonesia.

Francis, P. and James, R. (2003) Balancing rural poverty reduction and citizen participation: The contradictions of Uganda's decentralization program. *World Development*, 31, 325–337.

Furlong, P. and Marsh, D. (2010) A skin not a sweater: Ontology and epistemology in political science. In Marsh, D. and Stoker, G. (eds) *Theory and Methods in Political Science*, 3rd edn. Basingstoke: Macmillan.

Garnaut Review (2011) Update Paper 1: Weighing the costs and benefits of climate change action. Canberra: Garnaut Climate Change Review.

Geist, H. J. and Lambin, E. F. (2002) Proximate causes and underlying driving forces of tropical deforestation. *BioScience*, 52, 143–150.

Gerring, J. (2010) Causal mechanisms: Yes, but... *Comparative Political Studies*, 43, 1499–1526.

Gibson, C. C. and Lehoucq, F. E. (2003) The local politics of decentralized environmental policy in Guatemala. *The Journal of Environment and Development*, 12, 28–49.

Government Of Indonesia, (2011) Presidential Decree 61/2011 on National Action Plan to Reduce Greenhouse Gas Emissions, Indonesia.

Grbich, C. (2007) *Qualitative Data Analysis: An Introduction*. London: Sage.

Gregersen, H., Contreras-Hermosilla, A., White, A. and Phillips, L. (2004) Forest governance in federal systems: An overview of experiences and implications for decentralization: Work in progress. Bogor, Indonesia: Center for International Forestry Research.

Gregersen, H., Lakany, H. E., Karsenty, A. and White, A. (2010) *Does the Opportunity Cost Approach Indicate the Real Cost of REDD+? Rights and Realities of Paying for REDD+*. Washington, DC: Rights and Resources Initiative.

Grieg-Gran, M. (2000) Fiscal incentives for biodiversity conservation: The ICMS Ecologico in Brazil. *Discussion Paper 00-01*. London: International Institute for Environment and Development.

Grieg-Gran, M. (2008) *The Cost of Avoiding Deforestation: Update of the Report Prepared for the Stern Review of the Economics of Climate Change*. London: International Institute for Environment and Development.

Guba, E. and Lincoln, Y. (1994) Competing paradigm in qualitative research. In Denzin, N. and Lincoln, Y. (eds) *Handbook of Qualitative Research*. London: Sage.

Hanley, N. and Spash, C. (1993) *Cost-Benefit Analysis and the Environment*. Cheltenham: Edward Elgar Publishing.

Harrison, M. (2010) Valuing the future: The social discount rate in cost-benefit analysis. *Visiting Researcher Paper*. Canberra: Productivity Commission.

Herold, M. and Skutsch, M. (2009) Measurement, reporting and verification for REDD+: Objectives, capacities and institutions. In Angelsen, A. (ed.) *Realising REDD: National Strategies and Policy Options*. Bogor, Indonesia: Center for International Forestry Research.

Inman, R. P. and Rubinfeld, D. L. (1997) Rethinking federalism. *The Journal of Economic Perspectives*, 11, 43–64.

IPCC (2006) *IPCC Guidelines for National Greenhouse Gas Inventories. Volume 4: Agriculture, Forestry and Other Land Use*. Intergovernmental Panel on Climate Change (IPCC)., Kanagawa, Japan: IGES.

Irawan, S. and Tacconi, L. (2009) Reducing emissions from deforestation and forest degradation (REDD) and decentralized forest management. *International Forestry Review*, 11, 427–438.

Irawan, S., Tacconi, L. and Ring I. (2013) Stakeholders' incentives for land-use change and REDD+: The case of Indonesia. *Ecological Economics*, 87, 75–83.

Johns, T., Merry, F., Stickler, C., Nepstad, D., Laporte, N. and Goetz, S. (2008) A three-fund approach to incorporating government, public and private forest stewards into a REDD funding mechanism. *International Forestry Review*, 10, 458–464.

Kaiser, K., Hofman, B., Kadjatmiko and Suharnoko Sjahrir, B. (2006) Evaluating fiscal equalization in Indonesia. *World Bank Policy Research Working Papers*, 3911, 1–36.

Kartodihardjo, H. and Supriono, A. (2000) Dampak Pembangunan Sektoral terhadap Konversi dan Degradasi Hutan Alam: Kasus Pembangunan HTI dan Perkebunan di Indonesia. Bogor, Indonesia: Center for International Forestry Research.

Kindermann, G., Obersteiner, M., Sohngen, B., Sathaye, J., Andrasko, K. et al. (2008) Global cost estimates of reducing carbon emissions through avoided deforestation. *Proceedings of the National Academy of Sciences*, 105, 10302–10307.

King, D. (1984) *Fiscal Tiers: The Economics of Multi-Level Government*. London: HarperCollins.

Kingdon, J. (1995) *Agenda, Alternatives and Public Policies*, 2nd edn. New York: HarperCollins College Publishers.

Koh, L. P. and Wilcove, D. S. (2007) Cashing in palm oil for conservation. *Nature*, 448, 993–994.

Köllner, T., Schelske, O. and Seidl, I. (2002) Integrating biodiversity into intergovernmental fiscal transfers based on cantonal benchmarking: A Swiss case study. *Basic and Applied Ecology*, 3, 381–391.

Kumar, S. and Managi, S. (2009) Compensation for environmental services and intergovernmental fiscal transfers: The case of India. *Ecological Economics*, 68, 3052–3059.

Kunce, M. and Shogren, J. F. (2005) On interjurisdictional competition and environmental federalism. *Journal of Environmental Economics and Management*, 50, 212–224.

Larson, A. M. (2002) Natural resources and decentralization in Nicaragua: Are local governments up to the job? *World Development*, 30, 17–31.

Larson, A. M. (2003) Decentralisation and forest management in Latin America: Towards a working model. *Public Administration and Development*, 23, 211–226.

Larson, A. M. and Ribot, J. (2009) Lessons from forestry decentralisation. In Angelsen, A., Brockhaus, M. and Kanninen, M. (eds) *Realising REDD+: National Strategy and Policy Options*. Bogor, Indonesia: Center for International Forestry Research.

Larson, A. M. and Soto, F. (2008) Decentralization of natural resource governance regimes. *Annual Review of Environment and Resources*, 33, 213–239.

Levine, S. and White, P. E. (1961) Exchange as a conceptual framework for the study of interorganizational relationships. *Administrative Science Quarterly*, 5, 583–601.

Lewis, B. D. and Oosterman, A. (2009) The Impact of decentralization on subnational government fiscal slack in Indonesia. *Public Budgeting and Finance*, 29, 27–47.

Mahoney, J. (1999) Nominal, ordinal, and narrative appraisal in macrocausal analysis. *The American Journal of Sociology*, 104, 1154–1196.

Mahoney, J. (2000) Strategies of causal inference in small-N analysis. *Sociological Methods and Research*, 28, 387–424.

Mahoney, J. (2001) Review essay: Beyond correlational analysis – recent innovations in theory and method. *Sociological Forum*, 16, 575–593.

Margono, B. A., Potapov, P. V., Turubanova, S., Stolle, F. and Hansen, M. C. (2014) Primary forest cover loss in Indonesia over 2000–2012. *Nature Climate Change*, 4, 730–735.

Mascia, M. B., Brosius, J. P., Dobson, T. A., Forbes, B. C., Horowitz, L. et al. (2003) Conservation and the social sciences. *Conservation Biology*, 17, 649–650.

May, P. H., Veiga Neto, F., Denardin, V. and Loureiro, W. (2002) Using fiscal instruments to encourage conservation: Municipal

responses to the 'ecological' value-added tax in Parana′ and Minas Gerais, Brazil. In Pagiola, S., Bishop, J. and Landell-Mills, N. (eds) *Selling Forest Environmental Services: Market-based Mechanisms for Conservation and Development*. London: Earthscan.
Mccarthy, J. F. and Cramb, R. A. (2009) Policy narratives, landholder engagement, and oil palm expansion on the Malaysian and Indonesian frontiers. *Geographical Journal*, 175, 112–123.
Meridian Institute (2009) *Reducing Emissions from Deforestation and Forest Degradation (REDD): An Option Assessment Report*. Norway: Meridian Institute.
Milne, M. (1999) *Transaction Costs of Forest Carbon Projects*. Bogor, Indonesia: Center for International Forestry Research.
Ministry of Agriculture (2009) Keynote address. *MAKSI Conference*. Bogor, Indonesia, 24 November.
Ministry of Environment (2009) Republic of Indonesia, Indonesia's Second National Communication, 2009, Indonesia Second National Communication Under The United Nations Framework Convention on Climate Change (UNFCCC), Ministry of Environment, Jakarta.
Ministry of Finance (2008a) Anggaran Pendapatan Belanja Daerah. The Ministry of Finance, Jakarta.
Ministry of Finance (2012) APBD TA 2012. The Ministry of Finance, Jakarta.
Ministry of Finance (2009) Ministry of Finance Green Paper: Economic and Fiscal Policy Strategies for Climate Change Mitigation in Indonesia. Jakarta, The Ministry of Finance and Australia Indonesia Partnership.
Ministry of Finance (2013) Profil APBD TA 2012, Direktorat Jenderal Perimbangan Keuangan Direktorat Evaluasi Pendanaan dan Informasi Keuangan Daerah, Jakarta. Available at: http://www.djpk.kemenkeu.go.id/publikasi/apbd/163-profil-apbd-ta-2012.
Ministry of Forestry (2005) Operasi Hutan Lestari II, May 2005. Jakarta, The Ministry of Forestry.
Ministry of Forestry (2008a) Statistik Kehutanan 2008. Jakarta, The Ministry of Forestry.
Ministry of Forestry (2008b) Penghitungan Deforestasi. Jakarta, The Ministry of Forestry.
Ministry of Forestry (2008c) Rekalkulasi Penutupan Lahan Indonesia Tahun 2008. Jakarta, The Ministry of Forestry.

Ministry of Forestry (2013) Statistik Planologi Kehutanan 2012. Jakarta, Direktorat Jenderal Planologi Kehutanan, Ministry of Forestry of Indonesia.

Ministry of Forestry (2014) Kriteria calon areal IUPHHK-RE dalam hutan produksi, Direktur Bina Rencana Pemanfaatan dan Usaha Kawasan, Direktorat Jenderal Bina Usaha Kehutanan, Kementerian Kehutanan, Bogor. Available at: http://www.forda-mof.org//files/Puskonser.pdf.

Mollicone, D., Achard, F., Federici, S., Eva, H., Grassi, G. et al. (2007) An incentive mechanism for reducing emissions from conversion of intact and non-intact forests. *Climatic Change*, 83, 477–493.

Musgrave, R. (1959) *The Theory of Public Finance: A Study in Public Economy*. New York: McGraw-Hill.

Myer, E. (2007) Policies to reduce emissions from deforestation and degradation (REDD) in tropical forests: An examination of the issues facing the incorporation of REDD into market-based climate policies. *Resources for the Future Discussion Paper 07–50*. Washington, DC.

Myers, N., Mittermeier, R. A., Mittermeier, C. G., Da Fonseca, G. A. B. and Kent, J. (2000) Biodiversity hotspots for conservation priorities. *Nature*, 403, 853–858.

Nagendra, H. (2007) Drivers of reforestation in human-dominated forests. *Proceedings of the National Academy of Sciences*, 104, 15218–15223.

Naidoo, R., Balmford, A., Ferraro, P. J., Polasky, S., Ricketts, T. H. and Rouget, M. (2006) Integrating economic costs into conservation planning. *Trends in Ecology and Evolution*, 21, 681–687.

Naidoo, R. and Ricketts, T. H. (2006) Mapping the economic costs and benefits of conservation. *PLoS Biol*, 4, e360.

Nakhooda S., Caravani, A., Wenzel, A., and Schalatek, L. (2011) The evolving global climate finance architecture: Brief 2. Overseas Development Institute and Heinrich Böll Stiftung North America. Available at: http://www.odi.org.uk/resources/docs/7468.pdf.

Nawir, A. A., Murniati and Rumboko, L. (2007) Forest rehabilitation in Indonesia: Where to after three decades? Bogor, Indonesia: Center for International Forestry Research.

Nepstad, D., Soares–Filho, B. S., Merry, F., Lima, A., Moutinho, P. et al. (2009) The end of deforestation in the Brazilian Amazon. *Science*, 326, 1350–1351.

Nepstad, D., Boyd, W., Azevedo, A., Bezerra, T., Smid, B. J. et al. (2012) *Overview of Subnational Programs to Reduce Emissions from Deforestation and Forest Degradation (REDD). The Governors' Climate and Forests Task Force*. Palo Alto, CA: EPRI.

Oates, W. (1972) *Fiscal Federalism*. New York: Harcourt Brace Jovanovich.

Oates, W. E. (1999) An essay on fiscal federalism. *Journal of Economic Literature*, 37, 1120–1149.

Oates, W. E. (2001) A reconsideration of environmental federalism. *Resources for the Future Discussion Paper*. Available at: http://ideas.repec.org/p/rff/dpaper/dp-01-54.html.

Oliver, C. (1990) Determinants of interorganizational relationships: Integration and future directions. *The Academy of Management Review*, 15, 241–265.

Pagiola, S. and Bosquet, B. (2008) Estimating the costs of REDD at the country level. *MPRA Paper No. 18062*. Washington, DC, Forest Carbon Partnership – The World Bank.

Palm, C., Tomich, T., Van Noordwijk, M., Vosti, S., Gockowski, J. et al. (2004) Mitigating GHG emissions in the humid tropics: Case studies from the alternatives to slash-and-burn program (ASB). *Environment, Development and Sustainability*, 6, 145–162.

Palm, C. A., Woomer, P. L., Alegre, J., Arevalo, L., Castilla, C. et al. (1999) Carbon sequestration and trace gas emissions in slash and burn and alternative land uses in the humid tropics. *ASB Climate Change Working Group Final Report, Phase II*. Nairobi: ASB Coordination Office.

Parker, C., Mitchell, A., Trivedi, M. and Mardas, N. (2008) *The Little REDD book: A Guide to Governmental and Non-governmental Proposals to Reducing Emissions from Deforestation and Degradation*. Oxford: Global Canopy Programme.

Parsons, W. (1995) *Public Policy: An Introduction to the Theory and Practice of Policy Analysis*. Cheltenham: Edward Elgar Publishing.

Pattanayak, S. K., Wunder, S. and Ferraro, P. J. (2010) Show me the money: Do payments supply environmental services in developing countries? *Review of Environmental Economics and Policy*, 4, 254–274.

Pedroni, L., Dutschke, M., Streck, C. and Porrúa, M. (2009) Creating incentives for avoiding further deforestation: The nested approach. *Climate Policy*, 9, 207–220.

Perman, R., Ma, Y., Common, M., Maddison, D. and McGilvray, J. (2011) *Natural Resource and Environmental Economics*, 4th edn. London: Pearson Education Limited.

Pirard, R. (2008) Estimating opportunity costs of avoided deforestation (REDD): Application of a flexible stepwise approach to the Indonesian pulp sector. *International Forestry Review*, 10, 512–522.

Resosudarmo, I. A. P., Barr, C., Dermawan, A. and Mccarthy, J. F. (2006) Fiscal balancing and the redistribution of forest revenues. In Barr, C., Resosudarmo, I. A. P., Dermawan, A., Mccarthy, J. F., Moeliono, M. and Setiono, B. (eds) *Decentralization of Forest Administration in Indonesia: Implications for Forest Sustainability, Economic Development and Community Livelihoods*. Bogor, Indonesia: Center for International Forestry Research.

Riau Provincial Forestry Office (2013) Laporan Tahunan Dinas Kehutanan Provinsi Riau Tahun 2012, Dinas Kehutanan Provinsi Riau, Pekanbaru.

Ribot, J. C. (2003) Democratic decentralisation of natural resources: Institutional choice and discretionary power transfers in sub-Saharan Africa. *Public Administration and Development*, 23, 53–65.

Ribot, J. C., Agrawal, A. and Larson, A. M. (2006) Recentralizing while decentralizing: How national governments reappropriate forest resources. *World Development*, 34, 1864–1886.

Richards, K. R. and Stokes, C. (2004) A review of forest carbon sequestration cost studies: A dozen years of research. *Climatic Change*, 63, 1–48.

Ring, I. (2002) Ecological public functions and fiscal equalisation at the local level in Germany. *Ecological Economics*, 42, 415–427.

Ring, I. (2008a) Biodiversity governance: Adjusting local costs and global benefits. In Sikor, T. (ed.) *Public and Private in Natural Resource Governance: A False Dichotomy?* London: Earthscan.

Ring, I. (2008b) Compensating municipalities for protected areas: Fiscal transfers for biodiversity conservation in Saxony, Germany. *GAIA – Ecological Perspectives for Science and Society*, 17, 143–151.

Ring, I. (2008c) Integrating local ecological services into intergovernmental fiscal transfers: The case of the ecological ICMS in Brazil. *Land Use Policy*, 25, 485–497.

Ring, I., Drechsler, M., Van Teeffelen, A. J. A., Irawan, S. and Venter, O. (2010) Biodiversity conservation and climate mitigation: What role can economic instruments play? *Current Opinion in Environmental Sustainability*, 2, 50–58.

Rist, L., Feintrenie, L. and Levang, P. (2010) The livelihood impacts of oil palm: Smallholders in Indonesia. *Biodiversity and Conservation*, 19, 1009–1024.

Robson, C. (2002) *Real World Research: A Source for Social Scientists and Practitioner-Researchers*. Oxford: Backwell Publishing.

Rondinelli, D. A. (1990) Decentralization, territorial power and the state: A critical response. *Development and Change*, 21, 491–500.

Rötheli, E. (2007) *An Analysis of the Economic Implications of Developing Oil Palm Plantations on Deforested Land in Indonesia*. Gland, Switzerland: Worldwide Fund for Nature.

Sabatier, P. (1988) An advocacy coalition framework of policy change and the role of policy-oriented learning therein. *Policy Science*, 21, 129–168.

Sabatier, P. (1991) Toward better theories of the policy process. *Political Science and Politics*, 24, 147–156.

Sabatier, P. and Mazmanian, D. (1979) The conditions of effective implementation: A guide to accomplishing policy objectives. *Policy Analysis*, 5, 481–504.

Sabatier, P. and Mazmanian, D. (1980) The implementation of public policy: A framework of analysis. *Policy Studies Journal*, 8, 538–560.

Sandker, M., Suwarno, A. and Campbell, B. (2007) Will forests remain in the face of oil palm expansion? Simulating change in Malinau, Indonesia. *Ecology and Society*, 12, 37.

Santilli, M., Moutinho, P., Schwartzman, S., Nepstad, D., Curran, L. and Nobre, C. (2005) Tropical deforestation and the Kyoto Protocol. *Climatic Change*, 71, 267–276.

Santos, R., Ring, I., Antunes, P. and Clemente, P. (2012) Fiscal transfers for biodiversity conservation: The Portugal local finances law. *Land Use Policy*, 29, 261–273.

Santoso, I. (2008) Perjalanan Desentralisasi Pengurasan Sumberdaya Hutan Indonesia. Paper presented in the International Seminar *Ten Years Along: Decentralization, Land, and Natural Resources in Indonesia*, Universitas Atma Jaya Jakarta, 15–16 July.

Schlamadinger, B., Ciccarese, L., Dutschke, M., Fearnside, P. M., Brown, S. and Mudiyarso, D. (2005) Should we include avoidance of deforestation in the international response to climate change? In Moutinho, P. and Schwartzman, S. (eds) *Tropical Deforestation and Climate Change*. Washington, DC: Instituto de Pesquisa Ambiental da Amazônia.

Schneider, A. (2003) Decentralization: Conceptualization and measurement. *Studies in Comparative International Development (SCID)*, 38, 32–56.

Searle, B. (2007) Revenue sharing, natural resources and fiscal equalization. In Martinez-Vazquez, J. and Searle, B. (eds) *Fiscal Equalization*. New York: Springer.

Shah, A. (2006) A practitioner's guide to intergovernmental fiscal transfers. *World Bank Policy Research Working Paper 4039*. Washington, DC: The World Bank.

Sigman, H. (2005) Transboundary spillovers and decentralization of environmental policies. *Journal of Environmental Economics and Management*, 50, 82–101.

Singer, B. (2009) Indonesian forest-related policies: A multisectoral overview of public policies in Indonesia's forests since 1965. Ph.D. thesis. France, Institut d'Etudes Politiques and CIRAD.

Skocpol, T. (1979) *States and Social Revolutions: A Comparative Analysis of France, Russia, and China*. Cambridge: Cambridge University Press.

Smith, K. (2011) Discounting, risk and uncertainty in economic appraisals of climate change policy: Comparing Nordhaus, Garnaut and Stern. *Garnaut Climate Change Review – Update 2011*. Canberra: Department of Climate Change and Energy Efficiency.

Stern, N. (2006) *Stern Review on the Economics of Climate Change*. Cambridge: Cambridge University Press.

Strassburg, B., Turner, R. K., Fisher, B., Schaeffer, R. and Lovett, A. (2009) Reducing emissions from deforestation: The 'combined incentives' mechanism and empirical simulations. *Global Environmental Change*, 19, 265–278.

Subarudi and Dwiprabowo, H. (2007) Otonomi daerah bidang kehutanan: implementasi dan tantangan kebijakan perimbangan keuangan. Bogor, Indonesia: Center for International Forestry Research.

Susila, W. (2004) Contribution of oil palm industry to economic growth and poverty alleviation in Indonesia. *Jurnal Litbang Pertanian*, 23, 107–114.

Tacconi, L. (2007) Decentralization, forests and livelihoods: Theory and narrative. *Global Environmental Change*, 17, 338–348.

Tacconi, L., Mahanty, S. and Suich, H. (2010) Forests, payments for environmental services and livelihoods. In Tacconi, L., Mahanty, S. and Suich, H. (eds) *Payments for Environmental Services, Forest Conservation and Climate Change: Livelihoods in the REDD?* Cheltenham: Edward Elgar Publishing.

Tedjasukmana, J. (2007) Heroes of the environment: Barnabas Suebu. *Time*, 17 October. Available at: http://www.time.com/time/specials/2007/article/0,28804,1663317_1663319_1669895,00.html.

Telapak and EIA (2006) Behind the veneer: How Indonesia's last rainforests are being felled for flooring. Bogor, Indonesia: Telapak, and London: Environmental Investigation Agency.

Tiebout, C. M. (1956) A pure theory of local expenditure. *The Journal of Political Economy*, 64, 416–424.

Tropenbos International (2010) Improving governance, policy and institutional arrangements to reduce emissions from deforestation and degradation (REDD). Jakarta: Tropenbos International.

UNFCCC (2009a) Report on the expert meeting on methodological issues relating to reference emission levels and reference levels, held in Bonn from 1 to 10 June 2009, 1F4C CMCa/yS 2B0S0T9A /2009/2. Bonn: UNFCCC.

UNFCCC (2009b) Cost of implementing methodologies and monitoring systems relating to estimates of emissions from deforestation and forest degradation, the assessment of carbon stocks and greenhouse gas emissions from changes in forest cover, and the enhancement of forest carbon stocks. FCCC/TP/2009/1. Bonn: UNFCCC.

UNFCCC (2013) Decision 9/CP.19. Report of the Conference of the Parties on its nineteenth session, held in Warsaw from 11 to 23 November 2013. Bonn: UNFCCC.

Upton, S. (2009) The impact of migration on the people of Papua, Indonesia: A historical demographic analysis. Ph.D. thesis. Sydney: University of New South Wales.

Uryu, Y., Mott, C., Foead, N., Yulianto, K., Budiman, A. et al. (2008) *Deforestation, Forest Degradation, Biodiversity Loss and CO2 Emissions in Riau, Sumatra, Indonesia*. Jakarta: WWF Indonesia.

Usman, S., Mawardi, Ms., Poesoro, A., Suryahardi, A., Sampford, D. (2008) The specific allocation fund (DAK): Mechanisms and uses. Jakarta: SMERU.

Vatn, A. and Angelsen, A. (2009) Options for a national REDD+ architecture. In Angelsen, A., Brockhaus, M. and Kanninen, M. (eds) *Realizing REDD+: National Strategy and Policy Options*. Bogor, Indonesia: Center for International Forestry Research.

Venter, O., Meijaard, E., Possingham, H., Dennis, R., Sheil, D. et al. (2009) Carbon payments as a safeguard for threatened tropical mammals. *Conservation Letters*, 2, 123–129.

Wertz-Kanounnikoff, S. (2008) Estimating the costs of reducing forest emissions: A review of methods. Bogor, Indonesia: Center for International Forestry Research.

Williams, W. (1975) Implementation analysis and assessment. *Policy Analysis*, 1, 531–566.

Williamson, J., Karp, D. and Daphin, J. (1977) *The Research Craft: An Introduction to Social Science Methods*. Canada: Little, Brown and Company.

World Bank (2007a) *Indonesia and Climate Change: Current Status and Policies*. Jakarta: The World Bank.

World Bank (2007b) *Public Expenditure Review: Spending for Development, Making the Most of Indonesia's New Opportunities*. Jakarta: The World Bank.

Wunder, S. (2007) The efficiency of payments for environmental services in tropical conservation. *Conservation Biology*, 21, 10.

Wunder, S. (2010) Forest decentralization for REDD? A response to Sandbrook et al. *Oryx*, 44, 335–337.

Yanow, D. (2000) *Conducting Interpretive Policy Analysis*. Thousand Oaks, CA: Sage.

Yee, A. (1996) The causal effect of ideas on policies. *International Organization*, 50, 69–108.

Zen, Z., Barlow, C. and Gondowarsito, R. (2005) Oil palm in Indonesian socio–economic improvement: A review of options. *Working Papers in Trade and Development 2005-11*. Canberra: Research School of Pacific and Asian Studies, Australian National University.

Index